POISONED WELLS

THE DIRTY POLITICS OF AFRICAN OIL

Nicholas Shaxson

palgrave
macmillan

First published in hardcover in 2007 by PALGRAVE MACMILLAN™ 175
Fifth Avenue, New York, N.Y. 10010 and Houndmills, Basingstoke, Hampshire,
England RG21 6XS. Companies and representatives throughout the world.

PALGRAVE MACMILLAN is the global academic imprint of the Palgrave
Macmillan division of St. Martin's Press, LLC and of Palgrave Macmillan Ltd.
Macmillan® is a registered trademark in the United States, United Kingdom and
other countries. Palgrave is a registered trademark in the European Union and
other countries.

ISBN-13: 978-0-230-60532-9 paperback
ISBN-10: 0-230-60532-X paperback

Library of Congress Cataloging-in-Publication Data Shaxson, Nicholas.
   Poisoned wells: the dirty politics of African oil / Nicholas Shaxson.
      p.   cm.
   Includes bibliographical references and index.
   ISBN 1-4039-7194-3 (alk. Paper)
   ISBN 0-230-60532-X (paperback)
   1. Petroleum industry and trade—Political aspects—Africa.   2. Petroleum
industry and trade—Moral and ethical aspects—Africa.   I. Title.
HD9577.A2S53   2007
338.2'7282096—dc22

                                                                    2006049259

A catalogue record of the book is available from the British Library.

Design by Letra Libre, Inc.

First PALGRAVE MACMILLAN paperback edition: May 2008
10 9 8 7 6 5 4 3 2 1
Printed in the United States of America.

For EMMA.

CONTENTS

ACKNOWLEDGMENTS

I would like to single out a very small number of people, and one organization, for special thanks. The first is Alex Vines at Chatham House, whose ability to walk unscathed through ethical minefields has always astonished me, and who has helped and encouraged me in so many ways. The second is Antony Goldman, who persuaded me to make that first trip to Equatorial Guinea, and who has been a source of help and advice ever since. I would also like to thank the Open Society Institute for their generous and unconditional grant, without which it would not have been possible to write this book.

I will also always remember fondly the folk at the press centre in Luanda more than a decade ago: Sonia, Senhor Minvu, Vita, Rufino, Domingas, Ernesto, and Lila. I should also like to give special thanks to Tio Karl Maier, José de Oliveira, and Mário Paiva. And who could ever forget Chris Simpson?

Over the years, a number of other people have given up their free time to answer my foolish questions, and to help me out in other ways. They are far too numerous to mention here, but I would like to extend particular thanks to Antoine Lawson, the late Christine Messiant, Clotaire Hymboud, Patrice Yengo, Jon Walters, Ricardo Soares, Andrew Manley, Michela Wrong, Dino Mahtani, the staff at the Nigerian Institute for International Affairs, Gerhard Seibert, Stephen Ellis, William Wallis, Rafael Marques, Chris Heymans, Patrick Smith, Michael Holman, Simon Foot, Joseph Hurst-Croft, George Frynas, Ken Silverstein, Nick Chapman, Karolina Sutton, Ella Pearce, Clive Priddle, Barnaby Philips, Lara Pawson, Colin McLelland, and Zoe Eisenstein. Obviously, none of these people bear any responsibility for anything here that they might disagree with.

This book contains fewer than a hundred thousand words, but I could have written two million. This is just my personal selection of the facts; I am sorry for the huge amount I have left out. I apologize to Ian Gary, for hardly mentioning Chad, and to John Ryle, for not including a chapter on Sudan, which would, among other things, have granted me more space to write about China's growing influence in Africa—a subject to which I have not done proper justice here. Finally, I would like to grant an honorable mention to my fine parents, without whom none of this would have been possible.

Introduction

A PARADOX OF PLENTY IN THE NEW GULF

In 2006 *Energy Intelligence* published a ranking of oil companies that may have surprised some people.[1] ExxonMobil, which has a bigger market value than Wal-Mart and Microsoft,[2] was only thirteenth on this list—smaller than two African firms, and less than one-twentieth the size of a company called Aramco.

The ranking is not mistaken. Aramco is the Saudi Arabian state oil company, controlling more than 260 billion barrels of liquid oil reserves, while the two African firms are the Nigerian and Libyan state oil companies, with more than 20 billion barrels each. ExxonMobil has just 12 billion barrels, while BP, Chevron, Total, and Shell are smaller. If you rank the companies by market value, which excludes state oil companies whose shares are not traded, the privately owned companies like ExxonMobil rise to the top. Yet in the increasingly fraught global tussle for access to oil, the ranking that matters is the one with Aramco, not ExxonMobil, at the top.

Many people cherish a view that western oil companies are agents of imperialism, forcing weak Africans to accept dictates from London and Washington. This was true once. But when OPEC countries like Saudi Arabia and Iran began to flex their muscles in the 1970s, they wrested control over their oil decisively away from the Seven Sisters—the haughty

1

ancestors of today's ExxonMobil, Chevron, BP, and Shell. Since then Africa's rulers have more or less reversed the stranglehold. The bigger African producers now typically earn 70 to 90 percent of the value of their oil for their treasuries,[3] once development costs are paid off, leaving only a modest margin for the private oil companies. This share is the best measure of the balance of power today. White people and their companies no longer pull the strings in Africa.

Leaders of smaller countries do, it is true, sometimes still feel vulnerable. Equatorial Guinea's president Obiang Nguema, for example, was badly rattled when mercenaries tried to topple him in 2004. France used a secret system of extraordinary reach and power, fed by African oil, to keep a tight grip on its former African colonies for decades after independence. But this grasp is now fading. Now, as Chinese companies appear on the scene, western oil firms are growing more anxious and more obedient to oil-rich African rulers. When BP, responding to transparency campaigns, promised in 2001 to publish financial information about its oil operations in Angola, it was threatened with termination of its contract. BP jumped smartly back into line.

It is a time of high and rising anxiety for western oil companies and consumers. As China, Brazil, and other countries compete ever more aggressively, "friendly" territories like the North Sea or Texas are maturing and the global hunt for energy is shifting to harsher places. "The good Lord," U.S. vice president Dick Cheney once said, "didn't see fit to put oil and gas only where there are democratically elected regimes friendly to the United States."[4] As Russia wields its oil and gas as a political crowbar against foreign adversaries, as South American leaders nationalize energy assets, and as conflict roils the Middle East, the producer countries are growing bolder.

An alluring alternative global source of energy is Africa, which is more important to the West than most people know. In 2005, the United States imported more oil from this continent than it did from the Middle East, and it imported more from the Gulf of Guinea—where the West African coast bends south from Nigeria—than it did from Saudi Arabia and Kuwait combined. In less than a decade this new gulf will provide a quarter of U.S. oil imports,[5] and by 2008 ExxonMobil will be producing more oil in Angola than it does in the United States.[6] West African oil and gas tends to be light and sweet—ideal for refining into motor fuels—and it lies right on the shore, a straight sail across, or up, the Atlantic to American or European refineries. In the last five years Africa accounted for a quarter of all new oil discovered around the globe, and this share will rise.[7]

For years West Africa seemed free of the risks that accompany Middle Eastern oil: since it was first discovered here in the 1950s it has flowed freely through numerous civil wars and coups. The oil does not pass through risky chokepoints like the Suez Canal, and much of it lies offshore, out of reach of angry locals who for years showed little of the anti-western activism that roils the Middle East. This fed a cozy postcolonial complacency in the West: Big Oil, it was assumed, would endlessly dominate feeble African rulers who would be too busy scrambling for the cash to let supplies be interrupted.

For those people who did notice that the citizens of these oil zones seemed to be getting poorer and angrier, the answer was to send a few dollops of aid to tide the natives over until the oil money kick-started their economies.

This paternalistic view has now given way to concern—even alarm. One jolt came on September 11, 2001, when the West began to understand better Africa's value as a reliable alternative energy supplier. Another came after Chinese president Hu Jintao stepped off a plane in Gabon in 2004 and announced a new era of trade, aid, and friendship with Africa "without political strings."[8] Chinese and other Asian companies soon began aggressively to pursue West African oil assets, with fast and substantial success. In early 2006, Angola overtook Saudi Arabia to become China's biggest source of imported oil.[9]

Fresh blows have landed. In 2005 world oil prices rose above $50 a barrel for the first time when a Muslim Nigerian militant, Alhaji Mujahid Dokubo-Asari, threatened chaos in the oil industry. Militant attacks have since taken half a million barrels of oil a day out of world oil markets, spiking world oil prices up further in markets already jittery about Iran, Iraq, Lebanon, Venezuela, and much else. "We have acquired the technology to deploy heat-seeking rockets, and these shall be used against convoys of dubious politicians and deep-offshore facilities," a Nigerian militant group said. "If the Nigerian state and her collaborators really believe that [the oilfields] are truly protected, let them think again. Let them think hard."[10]

Yet almost 15 years tramping around Africa's oil zones have taught me that disruptions to oil supplies are not the only international threats. African rulers have a rising tide of money at their disposal, fit for mischief. The West is vulnerable to this money. You may not yet know much about President Omar Bongo of Gabon, but he has been a key to a huge, corrupt, and secret globalized system. Fed with African oil, this contributed powerfully to a terrible malaise in France today, with riots in the Paris suburbs and the rise of racist far-right politicians. Similar arrangements exist elsewhere in the West, in different forms, linked to African (or Middle East-

3

ern) oil. Most of us are hardly aware of them. Africa's oil is spreading poison deep into the fabric of the international financial system and the rich world's democracies. This book reveals the links that people are failing to see, and uncovers threats to which we in the West are almost completely blind.

West Africa's oil is also generating a mix of grievance, instability, and great wealth that is not unlike the noxious cocktail that motivated and financed the September 11, 2001, hijackers. Recently I noticed an article in an Angolan online newspaper referring to a study by the U.S. Council on Foreign Relations that mentions efforts by North Korea to secure uranium from Angola. What struck me were comments posted by Angolans beneath this story. "If it were possible, I would sell to Bin Laden," one wrote. "These American dogs need to understand!"[11] This does not reflect the common Angolan view, which is much friendlier. Yet the sentiment cannot safely be ignored either.

In a 1994 essay the American writer Robert Kaplan said that parts of Africa are reverting to the world of the Victorian atlas, as contact with the West increasingly takes place through coastal trading posts surrounded by a lawless interior that is once again becoming, as the writer Graham Greene observed, "blank" and "unexplored." Oil-soaked hybrids of Kaplan's vision are emerging along West Africa's coastline, with uncertain consequences for global security. In 1970, just before the oil boom, 19 million Nigerians lived below the poverty line. Now, nearly $400 billion in Nigerian oil earnings later, 90 million or more Nigerians live below that line, and their anger is mounting.[12] In the Niger Delta a deadly dynamic that emerged from local fury about oil activities is morphing into a huge international criminal enterprise. Angola and Congo-Brazzaville reeled in and out of conflict for years; Equatorial Guinea has one of the world's most repressive and corrupt governments. Gabon's oil is slowly running out, and it nurses not a cushion of savings but, instead, unpayable debts. The list of trouble goes on.

Oil and gas pay for the intelligence services and armies that keep the boiling anger at arm's length. It is no coincidence that Africa's four longest-serving leaders all come from these conflicted oil zones.[13] Ricardo Soares de Oliveira, one of the few western academics to have studied these issues in depth, calls sub-Saharan Africa's oil nations successful failed states: countries with strong leaders backed by sophisticated state-owned oil and gas companies and other islands of competence, surrounded by unusually tormented societies and AIDS-ravaged millions who shoulder the burdens of their states' failures. "The contrast between the heightened expectations of oil producers and their populations, on the one hand, and the dire impact of oil dependence on development, on the other," he wrote, "is one of the most dispiriting tales of postcolonial hope gone astray."[14]

The forgotten hinterlands and turbulent seas of invisible Africans that have emerged around the glittering shopping malls and yacht marinas in Lagos, Libreville, or Luanda should worry us all, because of the presence, in the vicinity of all this anger, of great wealth that circulates in unpredictable ways. The United States Defense Department is now paying unusual attention to this region, and western think tanks are scrambling to understand what the hell is going on in these crazy places. West Africa's oil nations are as troubled as the countries of the Middle East, but they are too often ignored, perhaps because West Africans have not blown anything up in New York.

Resources like oil and gas should be a blessing for countries that produce it. Norway and Britain seem to have done well out of their oilfields, but in Africa the record is different. Producing oil seems to be a bit like taking cocaine: if you are already healthy it might invigorate you, but if you are weak or sick, as many African countries are, it can do you serious harm. For most of the countries in this book, oil and gas account for over 90 percent of exports. Oil can also be a bit like heroin: the injection of cash from each cargo delivers a feeling of well-being, but the effect over time is addiction. Just as heroin addicts lose interest in work, health, family, and friends and focus increasingly on the next fix, so politicians in oil-dependent countries lose interest in their fellow citizens, as they try to get access to the free cash. Some countries, like Indonesia, have managed and even broken the addiction, but again the record in Africa is dismal.

You might think that because Africa is poor, oil money should help, even if the politicians steal most of it. The cash certainly abounds: West Africa's daily oil output is worth about as much as all development aid flowing to the world.[15] Today African producers are wallowing in cash, splurging on roads, railways, international airports, caviar, and Hummer jeeps—and inflation is falling. What is wrong with that?

One answer is that the impressive growth data are a mirage for most ordinary Africans. As the cash tumbles in, agriculture silently suffers, while the cost of living, and the poverty level, often rises. The long-term effect is deadlier. "Ten years from now, twenty years from now, oil will bring us ruin," OPEC's founder, the Venezuelan Juan Pablo Pérez Alfonzo, said during the 1970s oil boom. "We are drowning in the Devil's excrement."[16] His words were unfashionable then, but he was proved right: the euphoria was transient and left a hangover whose effects still reverberate. Academics talk of a Resource Curse, or a Paradox of Plenty, and research shows that oil-rich countries tend to grow more slowly than their peers do,[17] they are more authoritarian and conflict-ridden,[18] and they lie nearer the bottom of Transparency International's famous corruption perceptions index.

Oil rigs are alighting all along this stretch of Africa's western coastline like giant metal mosquitoes, standing on the skin of the earth on spindly legs and drilling down with steel proboscises to suck out the fluid that is the lifeblood of the world economy. Like the biting insects, the rigs can cause irritation around the site of extraction, disrupting local communities or polluting farmland. But it is this resource curse—the stealthier, time-delayed payload that accompanies the extraction, just like the malaria that real mosquitoes transmit—that is the real threat.

For decades development theorists could not get their heads around the counterintuitive idea of oil as a curse. It only began to enter the minds of a myopic western public in the mid 1990s, as television images of oil spills in the Niger Delta accompanied international outrage over the execution of the Nigerian activist Ken Saro-Wiwa. Yet this was cast mainly as an environmental issue: in early 1999, when I told a British friend about oil-related horrors that I had stumbled across in Angola, she asked me, "Why doesn't Greenpeace do something about it?" Only then did I see the extent of the western blindness. The big problems—of which environmental destruction is just an ugly symptom—are deeper and even more dangerous.

Now corruption is moving up the agenda of the world's development agencies. Left-wingers scream that Big Oil, corrupting and swindling innocent Africans, is to blame, and others say that the real problem is crooked African leaders, stealing the oil money.

Both sides are missing much of the point. Could the oil companies stop the resource curse if they put their backs into it? Are these African emirs all just clownish supergluttons, who could behave better if only they felt like it? Is it not possible that they take their jobs more seriously than that? In African homes the knee-jerk reaction to visitors is typically not one of greed, but of extreme, often embarrassing, generosity. How did this stunning contrast between greedy rulers and generous people come about? Oil helps frame the problem differently, and suggests some answers.

A funny thing that I have discovered is that for any given disaster in the oil zones, people will say, "This was not about oil. It was about something else." They will point to ethnic rivalries, for instance, or to the president's personality, as the root cause. But if you delve enough into the background of these calamities, you will eventually discern this furtive character, veiled in secrecy, tapping the protagonists quietly on their shoulders and whispering evilly into their ears. The character is not oil companies but oil itself—the corrupting, poisonous substance. It is a bit like how global warming generates heatwaves or thunderstorms: you cannot fully blame it for any given freak event, but the big pattern is unmistakable. Oil money is the first offender in wreaking this damage, but this book will identify another secret culprit, oil's wicked partner—I will come to that later—which is not only the scourge of Africa but of the wider world, twisting democracy and market economics violently into malign forms.

What is to be done? Oil companies and venal African rulers certainly have appalling crimes to answer for, but shouting at them is not working. The oil companies obey the African politicians, whose oil insulates them from pressure. If we in the West want to tackle these threats, much of the answer lies in making changes at home.

Our attention should be directed not so much toward oil companies and dictators (though we should not let them off the hook either), but toward the malefactors most directly responsible for the unfolding disaster. In this book I will identify some very different approaches for tackling a problem that should worry us all.

The aid theorist William Easterly remembers a time, not so long ago, when economists looking at tropical countries pretty much ignored politics. "It was only later," he said, "that we economists realized that government officials are people, too."[19] Economic theories that ignore human behavior break down in these African nations. So, to try to understand better what is going on in the oil zones, I have chosen to focus not on the countries themselves but instead on a handful of their strangest, most remarkable, and most infuriating citizens. Their stories will help expose Africa's oil for what it is: a threat to liberty, democracy, and free markets around the globe.

FELA KUTI

HOW THINGS FELL APART IN THE OIL BOOM

I first came across the Nigerian musician Fela Kuti at a London concert in the late 1980s, and the experience unsettled me. Not that I had no experience of Africa: I was born in Malawi in southern Africa, where my mother had worked as garden supervisor for President Kamuzu Banda and my father for Britain's overseas development effort as a specialist in tropical soils. My childhood memories are pleasant and even thrilling: dodging hippos in our little sailing boat and fishing for *chambo* on Lake Malawi; listening to hyenas laughing at the moon; and a hunter in my bedroom at night, waiting for a leopard that had killed our dog Judy and got into our rabbit hutch, and that my parents caught sunbathing on our compost heap (my sister and I were grounded until he shot it). But despite my African childhood, I didn't feel a connection to exotic, brash Nigeria.

When I saw Fela in concert I was a university student thrashing uncertainly in seas of political correctness, and his rude, sexually suggestive antics with female dancers, as I stood in the audience beside fearsome student feminists, made me squirm in my tender white-boy shoes. I found his music too harsh at first, but as the night wore on the hypnotic Afrobeat wormed into my head. It was intoxicating.

The late Fela Anikulapo-Kuti, the Black President, the King of Afrobeat, is probably Nigeria's most loved citizen of all time. He has been compared to Bob Marley, James Brown, Bruce Lee, Elvis Presley, Muhammad Ali, Mick Jagger, Bob Dylan, and many others—though none of these comparisons quite pins him down. This wiry musical Hercules and African spiritualist called himself Abami Edo—"the weird one"—and he would smoke turnip-size joints and put on white face paint before enthralling crowds with mesmerizing stagecraft, which he called his "underground spiritual game."

"Man, we walked in the room, and the smoke knocked us down," Bootsy Collins, the bassist for the American funk legend James Brown, said of Fela's band after touring Nigeria in 1970.[1] "When I heard that, it was like, 'Man, this is IT.' . . . We were telling them they're the funkiest cats we ever heard in our life. . . . I mean, this is the *James Brown band,* but we were totally wiped out. That was one trip I wouldn't trade for anything in the world."[2] The Brazilian musical superstar Gilberto Gil said that meeting Fela changed his life, and made him feel like a tree replanted.

Fela was also a sexual tornado who said he liked to enjoy the pleasures of at least two women per day. Journalists reported him emerging for interviews from his bedroom in his underpants, with semen still dribbling down[3]; and once he married all 27 of his dancers in a single ceremony. "Cocaine I stop am after I discover say e dey kill prick!"[4] Fela told one interviewer in the pidgin English that is widely spoken in his native Lagos. (He reverted to his favorite natural aphrodisiac N.N.G., Nigerian Natural Grass.) Fela had referred to women as "mattresses," and rejected condoms as un-African. "That thing, I don't touch it," he told one Nigerian interviewer. "Condom means Ko Do Mi. . . . Ko Do Mi is Yoruba for You Are not Fucking Me. See?"[5] When he died of AIDS in 1997, more than a million people came into the streets of Lagos to mourn his passing.

Fela taunted the politicians with songs like "Coffin for Head of State," "ITT (International Thief Thief)," and "VIP (Vagabonds In Power)," and he suffered endless court appearances, police beatings, and torture. During the 1970s, the decade when punk rock became the cutting edge of counterculture and protest in recession-prone Europe and America, and petrodollars began to cascade into the Nigerian treasury, Fela led a more deadly, desperate form of protest on behalf of ordinary Nigerians who were facing the emergence of a vicious, predatory new oil-rich elite.

I have chosen to write about Fela partly because he illustrates the indefatigable spirit of this remarkable, bellyaching country, which has two to three hundred ethnic groups and perhaps 130 million people, a sixth of Africa's population. Its religious divisions pit more Muslims than there are in any Arab country except Egypt against similar numbers of Christians and followers of African religions, forming what the Kenyan-born thinker

Ali Mazrui called the grand laboratory of the new triple heritage: a giant western, Islamic, and traditional African melting pot. These divisions complicate the question of how to share the oil money, which makes up more than 97 percent of Nigeria's exports today.[6]

Fela's long career, which reached its heights during the 1970s oil boom and the subsequent hangover of the 1980s, also offers essential context for this book. Around 2003 the world began to see a new oil price boom; to understand what this new bonanza means for Africa's energy producers we should look at the last boom. Fela's is a wholly African story; the British, French, Americans, and other foreigners who have helped shape Nigeria only get passing mention here. I will turn to *them* a bit later on.

Elderly Nigerians will tell you how much better life was in the early years after independence in 1960. "We still had a good civil service inherited from the British," remembers Philip Asiodu, who was Nigeria's top oil man during much of the boom. "It functioned well enough. Codes of conduct were followed, without any glaring corruption or destruction. They talked about people taking 5 percent, 10 percent; these were two or three ministers in each political party, as fundraisers, mostly not putting their hands in the till. They had a vision for running Nigeria properly. The 5 percent, the 10 percent—it did not distort the programs or affect the officials' authority."[7]

Even if this is too rosy a view of the past, a comment by the author Karl Maier illustrates how far Nigeria has fallen since then, nearly $400 billion of oil money later. "It is as if [Nigerians] live in a criminally mismanaged corporation," he wrote, "where the bosses are armed and have barricaded themselves inside the company safe."[8]

The award-winning Nigerian writer Chinua Achebe offers a clue that goes a long way to explain the trouble with Africa's oil nations. There is nothing basically wrong with the Nigerian character, land, or climate, he said. Then he penetrated right to the heart of Nigeria's predicament: "A normal sensible person will wait for his turn if he is sure that the shares will go round; if not, he might start a scramble."[9]

Nigeria is a bit like the queue that Achebe suggests. A functioning queue is really two queues: a physical one and a mental one. Disrupt the physical queue—by nudging a truck through it, say, or dousing it with a fire hose—and if the mental queue remains intact, then order will reemerge, in the same way that stable countries recover from economic shocks or terrorist attacks.

But there is a more damaging way to disrupt a queue: push in at the front. This assaults everyone's belief in it, and if it happens enough, the scrambling starts and it will collapse. There is then no easy way to rebuild it, no matter how much you shout.

Take the analogy further. Imagine the queue consists of, say, Americans, French, British, Chinese, and others (pick your prejudice), just as Nigeria is divided between Yorubas, Hausas, Igbos, Ijaws, and others. The pushing-in is getting worse, and you think you notice that the culprits are mostly from the group that you hate. Your faith in the queue will collapse faster, and you will hate the other group a bit more. Now imagine that your whole family has sent you to represent them in this line for food, which is too scarce to go around. You will jostle even more aggressively, the scramble will intensify, and the strongest and slipperiest characters, and those who can form alliances with their own people to thwart the other groups, will stand the best chance of getting to the front. This image helps me understand better the contrast between the hospitality and generosity you find in African homes, and the venality of many African rulers. Most corrupt people act like that only because they know everyone else is.

You see the scrambling and indiscipline everywhere: real lines degenerating into free-for-alls, or drivers forcing themselves selfishly into traffic and blocking everyone, are a version of this lack of trust and respect. Politicians think that if they don't grab what they can, someone worse will get it, so even if a dollar's worth of road repair saves a thousand dollars in broken axles, the holes remain unfilled.

As the citizens of this ethnically and religiously fractured nation have jostled, then scrambled, for what they can get of the oil money, long-term planning, shared nationhood and trust in each other—the keys to Nigeria's economic development—have dissolved. This splintering of the national good is really what corruption *is*. Later in the book I will propose a way of restoring some of this trust.

In the early years of independence, Fela studied at the Trinity College of Music in London. Another student remembered him as "a stunning extrovert [who] regularly held court among the bedazzled students, whom he often left speechless." Fela's landlady charged him for making extra noise with his trumpet, so he just made more noise. When the police were called, Fela called a London bobby a "foolish bastard." He was thrown into a police van,[10] but was soon back at his studies.

The British had handed over a country with three regions: the North, roughly corresponding to the hegemonies of the mostly Muslim Hausa-Fulani; the West, dominated by Yorubas; and the predominantly Igbo East. The North produced grains and groundnuts, the West grew cocoa, and the East produced palm oil. Each pulled its own weight in the federation, more or less, and this mitigated political tension. But oil, which had been discovered in 1956, was beginning to grow in importance. As it grew, the relationship changed fundamentally, as regions now had to compete for their share of the cake from an oil-fed center. Oil-producing areas said that they should get the most, the more populous regions argued that they should take precedence, while the poorest felt that they were most deserving. It was a huge, endless, unwinnable argument. Before long, regionally based political parties were clashing. Though the fighting was not obviously about oil, it was on everyone's minds.

Nigerians talk of a "National Question," which has many variants but is essentially this: how can the federation best be configured to hold its bickering groups together? Over time, in the spirit of old British divide-and-rule policies, Nigeria's rulers first split the regions into states, then split the states, again and again. Each new state had its own minorities who felt that the dominant groups in their state were snaffling the cash, so they pushed for their own smaller states, to get more of the cake for themselves. (Today there are 36 states.) Each new subdivision had new configurations of minorities, so the bickering continued, only in more decentralized ways.

By 1966, oil made up a third of Nigeria's exports,[11] and some Igbo officers, resentful that federal spending was skewed (they thought) to favor northerners, mounted a bloody coup, vowing to fight corruption, tribalism, and the enemies of progress: "the ten percenters, homosexuals, feudal lords, etc."[12] Many Nigerians saw this as an Igbo power grab, and an army general, Yakubu Gowon, launched a successful countercoup six months later. So Igbos in the East, hoping to carve out their own zone and grab the oilfields, decided to secede from Nigeria altogether.

In a last-ditch move to undermine eastern unity, the government split the federation into 12 states in 1967. But it was of no use: three days later the Igbo general Emeka Ojukwu declared an Independent Republic of Biafra, triggering civil war.

"It wasn't really a war for oil—I think it would have happened anyway," remembers Sola Odunfa, the veteran BBC reporter who accompanied Nigerian troops into Biafra for the *Daily Times*. "But oil encouraged them."

France hoped to wrest this oily gem away from the dreaded Anglo-Saxons, so President Charles de Gaulle sent the Biafran rebels planeloads of weapons, while Britain backed the Nigerian government. I remember being haunted for months by a vivid television clip from this war—I saw it while I was a teenager, quite a long time afterward—showing several federal soldiers manhandling a civilian, who was pleading for his life and repeating, "I am not a Biafran soldier! I am not a Biafran soldier!" They then pushed his head down into some bushes and shot him dead in front of the camera.

Perhaps a million Nigerians died in the three-year war, which remains today the most obvious example of oil's destabilizing push-pull effect: first, oil pushes Nigerians apart as they fight each other for the cash; then, it pulls them together again as they seek to remain connected to the oil-gorged federal center.

After the war, General Gowon promised a policy of "no victors, no vanquished" and ushered in a period of postwar optimism, as oil output was re-established.

By now Nigeria's leaders, fresh from military victory, angry at western support for apartheid South Africa, and marveling at Iran's efforts to wrest control of their industry from the companies (and also at the humbling of the American superpower in Vietnam), were feeling bold. At a time when it was fashionable for developing countries to "occupy the commanding heights" of their economies, Nigeria decided that the time had come to confront the secretive Seven Sisters, which they suspected used offshore accounting tricks to veil their real profits and cheat the Nigerian tax man.

"The Seven Sisters . . . were a monopoly, a cartel, and they effectively dictated the price of oil," the oil man Philip Asiodu said.[13] "The miracle of the Japanese and European recoveries after World War II was predicated on this cheap oil. The oil companies owned the assets, and we were aware that if the companies spent $100 million on procurement, they spent $95 million of that outside Nigeria. Ours were legitimate dreams."

First, Nigeria ramped up the price of its oil and raised tax rates, then increased audit requirements on the companies and even made them incorporate themselves locally, so that they were subject to Nigerian jurisdiction.

The companies closed ranks, hoping to form a united front to resist the African upstarts. But the world was changing. Charles de Gaulle's foolish support for Biafran rebels meant that French oil interests were reduced to pleading, and to accept humiliating new terms. And the maverick Italian oilman Enrico Mattei, who hated the *Sette Sorrelle* (Seven Sisters—a term that Mattei is credited with coining) had, before he died, presented his group, ENI, as being a more flexible, fair-minded alternative. ENI had already offered the Iranians three-quarters of the revenues from production in Iran, instead of the half that the Sisters were paying, and it was happy to accept new terms in Nigeria, too.

The Sisters' united front was crumbling. Nigeria unilaterally took over a 60 percent controlling stake in several of their large exploration and production licenses. The companies, Asiodu said, lobbied ferociously against this, and even got military officials to apply pressure on him to stop the reforms. "I told the companies that it was not our philosophy to make them have a loss, and I would quite understand it if they wanted stop operating in Nigeria," he said with a chuckle. "That was not what they wanted to hear."

Nigeria's oil production was soaring, too, having risen from 150,000 barrels per day in 1968 to one and a half million by 1971, then exceeding two million in 1973: more than twice as much as the United States was then importing from the Persian Gulf. [14]

The stream of petrodollars became a river. Then, with the Yom Kippur war and the OPEC oil embargo of 1973, it became a tide as world oil prices quadrupled from $3 to $12 per barrel in just three months. (Later, with the fall of the Shah of Iran in 1979 and the Iran-Iraq war, prices rose to nearly $40,[15] worth nearly $100 in today's prices.[16]) The effect on Nigeria was staggering: by 1975 oil made up 95 percent of exports, and between 1970 and 1980 its annual export earnings rose from 1 billion dollars to 26 billion.[17]

It changed everything. The politicians promised to harness the oil in a great leap forward,[18] and well-connected Nigerians scrambled for government contracts to build bridges, flyovers, railways, and so on, in an orgy of post-war construction. The defense ministry handed out a rustling torrent of cement import contracts, and soon a fabled armada of ships arrived, loaded with the world's cement. At one point in 1975, 400 ships were waiting, collecting demurrage charges. Some even registered with the Nigerian authorities, then sailed elsewhere on business while collecting the demurrage payments. Nigeria erected import tariffs to encourage local manufacturing, which did not make local industry more competitive, but instead generated profits for tricksters. In a famous moment of frankness, the military ruler General Gowon conceded that Nigeria's problem was not the money, but how to spend it.

Salaries went up. "The whole place was awash with cash," the BBC's Sola Odunfa remembers. "People got nine months' salary arrears paid. Oh booooooy! Shops were full. People were buying radios, fridges; you saw people on motorbikes with televisions on their heads. Just crazy. Everyone had money and wanted to spend. Champagne, watches, cars, or—if they were too poor—just motorcycles. There was no time to think; it was spending, without productivity. We did not yet see that boom would become doom. Some say that era destroyed us. The corruption—it felt like it exploded from nowhere."

Suddenly, General Gowon was toppled in another coup, ostensibly because he had decided to delay elections. But oil was, once again, the silent player. Ahead of the coup, as everyone jostled for contracts, tension grew between the oil ministry and the national oil company about who controlled the oil sector, and this was stirred by foreign traders and Nigerian middlemen fighting for access to the oil money. Matters came to a head in a government meeting in July 1975, in which the oil man Philip Asiodu recommended one of his protégés as general manager of the national oil company,[19] to the fury of Murtala Muhammed, a respected army general. General Gowon rejected Muhammed's advice, and 25 days later Muhammed mounted the coup. The oil appointment, he later admitted, was "the straw that broke the camel's back."[20]

Asiodu remembers what happened next. "Over six weeks, without due process, without controls, more than 10,000 civil servants were retired or dismissed," he said. "The justification was to shake up the civil service, which was too powerful. But who was shaking it up? Military officers without training. It was a disaster, it destroyed morale; it gave rise to the 'make hay while the sun shines' mentality: which is corruption. Our pride, as custodians of the public good, was destroyed. The Nigerian psyche has been assaulted and insulted. With civil servants respected, we would have been an African lion today beside Asian tigers. But now we are 45 years behind."

The new leader, Muhammed, was killed a few months later in another messy, abortive coup attempt in February 1976. Days afterward, the *New York Times* Nigeria correspondent John Darnton, who had just arrived, saw a bizarre caravan of young people in the street, led by a Day-Glo bus. "What's that?" he asked. "That's Fela," someone said. "To the government, he's nothing but trouble."

Darnton resolved to meet the man. Whatever it was he expected, that wasn't what he got.

It's not always easy to realize when you're in the presence of a genius, especially when it comes in the form of a muscular five-foot-seven Nigerian, dressed in leopard-skin bikini underpants, his eyes blurry red from overindulgence in marijuana. He was seated on a thronelike chair . . . smoking a cigar-size joint that was held for him between tokes by one of three or four female attendants. The interview was awkward at first, but he soon warmed up; he was grateful to America, which he had visited in 1969, for teaching him about Black Power, he said. It was odd, he added, but it took photos of African-Americans wearing dashikis on 125th Street for Nigerians to feel proud in their own national dress. What he most disparaged about the United States was the size of the joints: "Do you believe," he told his circle of wide-eyed followers, "over there, they light up one little one, and they have to pass it around!"

Later, on the street, Fela stopped traffic for blocks around, followed by crowds who chanted his name, throwing clenched-fist power salutes.[21]

M. D. Yusufu, Nigeria's top policeman at the time, remembers Fela well, and worked hard to befriend him. "Even though I was chief of police, I realized that Fela was the one person who could come out with a demonstration and block the two entrances to Lagos and stop everything," he said. "His youths go everywhere. I thought that as a policeman I'd better be friends with that one, despite his lewd behavior."[22] Yusufu recalled a visit to Lagos of the Guinean leader Sekou Touré, who invited Fela for a meeting. Fela arrived with around 25 wives, and there was a scuffle with police outside before he got in. "Sekou Touré said to Obasanjo, 'I have such people in my country, but long ago I realized that they are very dangerous opponents. So I became patron of all the artists, for the simple reason that they have a talent that no government has. In one song, he can destroy years of my work.'"

The coup that Darnton had arrived for failed, but General Muhammed had died in it, so his deputy, General Olusegun Obasanjo, reluctantly took over.

As the boom progressed, oil had another curious effect: the collapse of agriculture. One problem was that millions of rural Nigerians downed hoes and, like moths to an oil lamp, flocked to the towns and cities. Meanwhile, civil servants lost interest in agriculture as they focused instead on wheedling more money from the oil-fed budget.

But another thing was happening too: the dreaded Dutch disease. This generic effect was named after economic disasters that hit the Netherlands, where I now live, after it made big gas discoveries in the 1960s. The windfalls bring cash cascading in, raising price levels and so making local goods more expensive and less competitive when compared to imported goods. Sectors like industry and agriculture wither, and oil pays for a rising tide of imports to replace their output. In the process, a relatively small number of people get rich, and farmers are left in the dirt.

Just before the boom, Nigeria had been the world's second producer of cocoa,[23] and agriculture made up three-quarters of exports.[24] Between 1975 and 1978 alone the total area under active cultivation fell by 60 percent[25] and recorded production of major crops fell by about as much,[26] plunging tens of millions into poverty. It was a classic display of oil's jealous character, crowding out and crushing other economic sectors.

Something else was underway too. A rallying cry of the American revolutionary war against London was, "No taxation without representation"—people will pay taxes only if they feel that they get a political voice in return. But in Nigeria, as Fela quickly noticed, the rulers no longer needed their citizens for their tax revenues, since they had the oil money instead. Millions of Nigerians might as well have fallen off the map.

Fela's early songs reflected the exuberance of the times and his own inner mischief. His dazzling dance-floor anthem "Jeun K'oku" emerged in 1970, just as the Biafra war ended, and it entranced Nigerian youths more used to the Beatles, Simon and Garfunkel, Otis Redding, or West African Juju or Highlife. " 'Jeun K'oku' came and blew us away," a Nigerian academic remembers of that time.[27] "It was local, it spoke about our home reality, and it met our critical requirements of sophisticated sound."

His sultry 1972 hit "Na Poi," making cheeky and explicit references to the sex act, rippled anger across social, religious, and official strata—which, of course, just made Fela more popular. "Male students on campus," wrote one Nigerian journalist, "could not sing it loud enough."[28] Fela also moved away from singing in his native Yoruba tongue toward the more widely-spoken pidgin English, boosting his audience.

His songs also began to take on a more political edge. Like the politicians standing up to the Seven Sisters, Fela was growing bolder, too. On a visit to the United States he had dated Sandra Iszidore, a former member of the Black Panthers who had spent time in jail for assaulting a police officer during the Los Angeles riots. She turned Fela on to the hedonistic, drug taking, sexually liberated American counterculture, and tuned him in to the Black Power movement and the ideas of thinkers like Malcolm X and Marcus Garvey. "For the first time, I heard things I'd never heard before about Africa," Fela later said. "She was the one who opened my eyes."[29]

These ideas intertwined in his mind with pan-African ideals that he had imbibed from his fiery mother, a leading agitator against British colonial rule who had often thrashed Fela as a child.

He hung pictures of Kwame Nkrumah and other pan-African heroes at the Afrika Shrine in Lagos, the chaotic musical commune where he proclaimed himself chief priest and where traditional leaders offered libations as Fela worshiped his ancestors. (One visitor said the Shrine looked like a cross between a Black Panther safe house and the Playboy mansion.) It became the center of West Africa's music scene, and Motown even offered Fela a million-dollar deal (despite Fela's tendency to play hour-long songs, and never to play old material.) The deal fell through after Fela's personal spirit medium, Professor Hindu, advised him not to sign.

The ex-Beatle Paul McCartney, who visited Lagos in 1972, said that Fela's group was "the best band I've ever seen live. . . . When Fela and his band eventually began to play, after a long, crazy buildup, I just couldn't stop weeping with joy."[30] McCartney wanted to use African musicians for the album *Band on the Run* that he was recording in Lagos, but Fela denounced McCartney, berating him for "stealing Black Man's music." Fearlessness and love of controversy were Fela's strength, but they also contributed to what a music journalist called his "unerring ability to piss on his own parade,"[31] perhaps explaining why Fela never finally became a true international superstar.

Yet at home, he was unsurpassed. He would strut onstage in his underpants, taunting Nigeria's elites with songs like "Gentleman," which he released in 1973, mocking members of the Nigerian elite who would put on ties, coats, and hats and end up sweating all over, and smelling "like shit":

I no be gentleman like thaaaat, I be Africa man, original.

In 1975 Fela changed his "slave" name from Ransome-Kuti to Anikulapo ("one who carries death in his pouch") Kuti, and renamed his compound the Kalakuta Republic, surrounding it with barbed wire and declaring it an autonomous zone free from Nigerian law. Soldiers raided it constantly. In 1977, when President Obasanjo's government splurged over 100 million petrodollars on an international jamboree called Festac, Fela organized a Counter-FESTAC at Kalakuta, complete with dancers and buckets of grass. Stevie Wonder, other stars, and the crowds flocked to see Fela. This incensed the military government, which was already furious about his recent smash hit song "Zombie." It portrayed soldiers as mindless robots who would not even think unless they were told to do so, and it incited street youths to brandish mock rifles and taunt them in the streets.

Go and die—joro-jara-joro [left, right, left], No brains . . . no sense.

Days after Festac, hundreds of soldiers descended on Kalakuta, in a frenzy. They smashed testicles with rifle butts, they dragged women naked to army barracks and tortured them with bottles, and burned Kalakuta down. Fela was hauled out by his genitals and suffered a cracked skull; his brother Beko was in a wheelchair for months; they threw his 78-year-old mother through a window. Fatai Rolling Dollar, an aging, impish Nigerian musician who lived down the street and who also was injured in the attack, described—amid clouds of pungent smoke after a show of his that I saw in Amsterdam in 2006—how he lost all his instruments and spent the next quarter century destitute, doing security guard work and other menial jobs.

Fela's son Femi, then age 14, later recounted to me his painful memories of what he saw, coming back from school. "Soldiers in full gear, battle gear. People were scared, screaming, so we thought he was dead. We found him in the barracks, in the military hospital, about 12 midnight. They let us see him for, like, two minutes. He told us he was all right. He had been beaten, but he told us he was OK, and we just gave him a hug and then left."

In prison, Fela was a hero, and when he got out, many months later, he was as brassy as ever, composing "Unknown Soldier," referring to the military inquiry's explanation for who destroyed Kalakuta, and another, a more painful song that he composed a while after the attack, after his elderly mother had died of her injuries. The song, "Coffin for Head of State," described how he had carried his mother's coffin and delivered it in person to Dodan army barracks, putting it down before the gate.

Obasanjo dey there, with him big fat stomach, dem no want take am.

The oil politics found strange echoes in Fela's life. While Nigeria's leaders bickered over how to share the oil money, Fela's entourage was embroiled in something similar: the old story of a rock band tearing itself apart as its members argue over how to divide the spoils and the glory. Tony Allen, Fela's inspirational drummer, remembers the easy money of the 1970s, when Nigerians could afford to flock to their shows. But by the end of the decade Allen was fed up, and left; he called Fela a "slave driver" in one interview. Years later, Allen was more circumspect when I met him. "He was a genius," he said. "But I had had enough of too many imbeciles: people I didn't fucking know from anywhere, who had not been around when we were struggling. As soon as we had arrived, we saw all these bees. Bzzz, bzzz, they overtook everything. Fela accommodated them," he said. "They were just yes-men, riffraff, guys that go out, come back with unnecessary things, fictitious stories, just to amuse the guy, to make him react to what he never saw, to build their own standing. The money was coming in, more and more. This was power, money, the thing was terrible."

In 1979, President Obasanjo handed over power to civilians (the Nigerian people would later thank him, reelecting him in 1999). It was also the year that the Shah of Iran fell and oil prices spiked again. Obasanjo said that Nigeria would be among "the ten leading nations in the world by the end of the century."[32] It was not, of course, to be. Producing oil is not like producing shoes or bicycles. Oil projects are big and complex, and crude oil often does not start flowing until many years after a first discovery; a rise in prices takes a long time before it stimulates more exploration and a rise in supply. It can also take years for higher prices to work their way through the world economy and choke off demand. In the meantime, the supply-demand balance can shift far out of kilter, meaning that prices rise much too far. When supply finally rises and demand finally subsides, prices can also fall dramatically, as the balance tips too far the other way.

This is what happened when the oil boom decade ended: world oil prices fell from nearly $40 a barrel in 1980 to just over $10 five years later—a collapse that was even more vicious when you take the high inflation of those years into account. For Nigeria this was aggravated by oil contracts that, for historical and technical reasons, exaggerated the effects of price fluctuations on government revenues.[33] From $25 billion in 1980, Nigeria's oil export earnings fell to $10 billion in 1983, and just $7 billion by 1986.[34] For a government whose credibility by now depended on doling out oil money, this was bad enough. But also, problems that had quietly built up in the boom did not unwind peacefully, but instead mutated into new, more pernicious forms.

As spending fell, each rival faction scrambled to feed itself, at the expense of other factions and of the more ephemeral long-term goal of national development. The factions were not only ethnic and religious: the civil service was a faction, as were political parties, state governments, local authorities—which all fought to maximize their shares. Faction leaders were often fat cats milking oil-subsidized government projects, which had been easy to set up during the oil boom, but politically much harder to shut down after the bust. Of course the strongest and slipperiest usually won and many of the cash-cow projects kept going, while Nigeria's poor bore the brunt of cuts, and austerity programs encouraged by IMF-inspired belt-tightening. The IMF became a focus of hatred. Mark Allen, who visited Nigeria for the IMF in the 1980s, remembered a soldier saying he would put a bullet between the eyes of IMF officials if he ever got hold of them. "One mission they gave us a bullet-proof car with glass this thick," Allen said, measuring an inch and a half with his fingers, "and a driver trained in the U.S. who knew how to drive away at 70 mph with the tyres blown out."[35]

As millions sank deeper into poverty, a new "austerity bourgeoisie" got rich. They were local and foreign middlemen and politicians who got the bureaucrats to dream up price controls and other economic distortions for them to exploit. "Somehow the oil money seemed to be getting lost very fast in the black box of the oil ministry, somewhere between the oil companies paying it and the finance ministry receiving it," the IMF's Allen said. "They understood our questions; they showed equal enthusiasm to get to the bottom of it all—then there would be long rigmaroles. . . . We never got very far."

Nigeria's parliament has long been especially good at teasing out the money, and Allen contrasts it with the British parliament, which was set up to scrutinize and control the King's spending requests. "Nigeria's parliament is more like the U.S. Congress with its pork-barrel projects: [its] job is not to protect the money but to get hold of it. They are huge spending lobbies."

Agriculture recovered a bit as the oil revenues collapsed in the 1980s, but this time the main beneficiaries were not rural farmers but big farmer-capitalists.[36] Conditions worsened, deepening and widening Nigeria's fracture.

The period of civilian rule after 1979, not the military rule that preceded it, was the time when economic discipline really deteriorated, and when Nigeria's foreign debt grew fastest.[37] The new civilian rulers, lacking military backbone, felt more need to shore up their political support by doling out contracts and favors, and even as oil prices fell they spent recklessly: building a new national capital in Abuja and pushing ahead with the giant Ajaokuta steel project, a classic white elephant that was supposed to be Africa's biggest steel plant yet ended up consuming billions of dollars without producing any steel.[38]

Meanwhile, Muslims and Christians in the North fought bloody battles over sharia (Islamic) law, which offered quick, clear and efficient alter-
natives to the state's crumbling legal institutions. Oil was never the primary
focus of feuds, but it lay in the background, as usual. Religion, like tribalism, was useful for local leaders wanting to mobilize people, often violently,
to back their claims on the petrodollars. The media focused on the violence, but this was just the most visible part of more dangerous conflicts
churning out of view, inside the government. As oil sharpened the competition, moderate religions became radicalized.

Fela called the new democratic dispensation Demo-Crazy, Demon-
Crazy, and Dem-All-Crazy. The civilians' cup of iniquity filled up fast, and
when northern soldiers mounted another coup in 1983, many Nigerians
were relieved. The new leaders launched a "War against Indiscipline," even
sending soldiers into the streets with whips to make people wait in orderly
lines at bus stops, and getting civil servants who were late for work to do
humiliating public "frog jumps." The leaders even tried to recover some of
the funds stolen under the civilians, but Britain and other western governments and banks, nervous about getting a reputation among dictators and
crooks for not protecting their lucrative cash deposits from foreign investigations, disgracefully declined to help.[39]

But the soldiers became more authoritarian, bulldozing informal set-
tlements and markets, and imprisoning many people. Fela's "Beasts of No
Nation," which he wrote in jail while serving time for foreign currency vio-
lations, captures the spirit of the time. The album cover depicts Margaret
Thatcher, Ronald Reagan, and South African apartheid president P. W.
Botha with devil horns and blood dripping from their mouths.

Even as the oil revenues collapsed, the factions kept guzzling cash, so
yawning deficits opened up, and foreign debt grew.[40] In the boom years
Nigeria had found it easy to borrow money (bankers love lending to rich
people, and in the 1970s their coffers were stuffed with Arab petrodollars),
and foreign debt had reached $5 billion by 1980. Now Nigeria fell into ar-
rears, and the penalty payments meant that its debt grew even more: by the
end of the decade Nigeria was nearly $30 billion in hock.[41] It was a classic
boom-bust debt trap, which was happening in oil-producing countries all
around the world.

Since then the mayhem has continued. The coups, and the constant
pushing and pulling from oil, have weakened Nigerians' sense of law and
trust in each other. When Saddam Hussein invaded Kuwait and oil prices
spiked again, over $5 billion disappeared from government accounts; some

Nigerians said that the reign of General Ibrahim Babangida from 1985 to 1993 was the most corrupt of all to date. He was followed by the brutal dictator Sani Abacha, who was even more reviled until his death in 1998 from poisoning, in the arms of Indian prostitutes. After that, while a military caretaker ruler[42] paved the way for elections, huge amounts of money were sucked out of the treasury and stashed offshore.

Between 1970, when the oil boom started, and 2000, while Nigeria earned more than $350 billion from its oil, the poverty rate rose from 35 percent to 70 percent, as incomes steadily grew more unequal and as the economy shrank.[43] In the 1960s Nigerians and Indonesians had roughly the same average income, and their countries have since earned similar amounts from oil.[44] Yet Indonesia's economy has quadrupled, while Nigeria's has shrunk. There are many complex differences between the countries, and one simple one. Indonesia's long-standing President Suharto was an authoritarian, who for all his failings at least promoted a unified vision for his country. But in Nigeria no group ever dominated.

I'm not saying that authoritarianism is good; just that in fragile countries strong central governments backed by a political consensus, usually coming from a strong middle class, tend to perform better than weak, fractured ones. (Oil, by magnifying inequality and attacking the industrialized parts of the economy, often destroys the middle class.) Academic research has found that divided societies tend to be more violent, and grow more slowly, than others.[45]

"In a society when everyone cheats and takes or pays bribes, there is little incentive not to join in," wrote the *Financial Times* commentator Martin Wolf. "Government is a monopoly for good reason. Competing bandits are bad news."

Today Nigeria has a civilian government under President Obasanjo, who was elected in 1999 as a paragon of transparency and good governance. He cleverly appointed Nigerians recruited from the overseas diaspora—less suffused with the Nigerian scramble mentality—and placed them in a self-reinforcing ring, where they would hopefully be able to trust each other even if they couldn't trust anyone else. They boast remarkable achievements: instead of spending the new oil windfall chaotically, they have instead built up big savings; and agreed with Nigeria's creditors in 2005 to pay off a most of Nigeria's vast foreign debt.[46] Yet despite this, many Nigerians feel that their country has plumbed yet new depths, as Obasanjo has made compromise after compromise in order to shore up his political support base. In a survey in 2006, 78 percent of Nigerians said they thought corruption had increased during Obasanjo's rule.[47]

Allison Ayida, Nigeria's top finance official for much of the 1960s and 1970s, is now a creaking, elderly gentleman who totters around a large, comfortable house in Lagos. When we met, my tough, brash driver bent down in front of him, touching the ground by his feet with his hands. Ayida ushered me into a large room and closed the door to keep children from bouncing in while we spoke. As he reminisced about how Nigeria has changed, he slowly shook his head: "This is the same country I knew before. It was a strict system before; you were held accountable. The legislature was taken seriously. But the controls broke down," he said. "The [civil servants] were watchdogs; now they are part of the looting. The state governments have to pay finance officials in order to get their allocation released. It is happening before my eyes today. I—I can't believe it."

Fela, perhaps sapped by all the beatings, grew more introspective after the 1980s. His commune degenerated and he divorced most of his wives. In the 1990s he lost weight and his health deteriorated, but he refused to be tested for why he had developed a persistent cough and skin lesions. He still called AIDS a "whiteman's disease," and continued to have unprotected sex. He died in 1997 and was buried in his tight yellow trousers, with a joint between his fingers.

"When my mother died, it was because she finished her time on earth. I know that when I die I'll see her again, so how can I fear death?" he had told an interviewer before he died. "So what is this motherfucking world about? I will do my part . . . then I'll just go, man. Just go!"[48]

Some Nigerians say that foreign aid money has had a hand in the degradation, making activists shift their attention away from real issues toward competing for donors' cash, in an effect echoing what Nigeria's oil does to its politicians. "In the past we had social movements, a lot of networking; their strength lay in their power of organizing," said Otive Igbuzor, Action Aid's country director in Nigeria. "Donor money has had a lot of negative impact on organizing. They used to meet in classrooms; now they meet in five-star hotels, and their motives for organizing are the per diems that donors pay them."

Nigeria's best-known activists—Fela, his late brother Beko Ransome-Kuti, the Ogoni activist Ken Saro-Wiwa, writers like Chinua Achebe or the Nobel prize-winner Wole Soyinka (who is Fela's cousin)—mostly hail from past generations. There is nobody of this stature in Nigeria today; Nigerians have become so polarized by their endless competition for resources, fragmenting the very issues themselves, that it is hard for anyone

25

now to speak for the whole nation. The loudest alternative voices come from armed militants who attack oil rigs in the name of their ethnic heroes, often against the very idea of a united Nigeria. Their shocking tales, revealing seething and unpredictable new global threats, will appear later in this book.

Fela's story, the musical score to accompany the unraveling of a nation, remains an essentially African tale. But it is important to remember that the dangerous effects of Africa's oil are not confined to this continent: its poisonous effects spread far overseas, to tangle secretly with a globalized financial architecture and with the shadow worlds of western politics. I will turn to this bizarre, invisible world in a little while.

Today, Fela's son Femi Kuti puts on magnificent performances at the new Africa Shrine in Lagos, attended by fanatic followers and a band of devoted cripples who slide around violently in front of the stage on makeshift skateboards. His song "Sorry Sorry," a lament for Nigeria, is enough to make people cry. Femi looks like a cleaner-cut version of his father, although he rarely sings Fela's songs, and he has fallen out with other family members, partly over disagreements as to who should get Fela's royalties. Femi had to buy the land for his new shrine, after rich landowners sued in court to appropriate Fela's original shrine.

In his chaotic dressing room behind the stage, Femi rolled joint after joint after joint, lining them up neatly in a large circular tin. He didn't want to talk much about his father, who he said died disillusioned, having spent his life fighting a system that just got stronger.

It is very, very, very boring. Truly, sometimes I don't see the importance of interviews, so I just stick to writing my songs. People have been talking about this for years and it has not seemed to change anything. Compare this to the 1970s—eeee, eeee, it has all been downhill since then. I have been talking about this all my life, my father talked about it, my mother talked about it, my son will probably talk about it. People don't complain because they are so weak. Weakened by the bitterness of this corruption. Co—rrrrr—uption! Nigeria, politically, it's a write-off. Everyone knows it. It is a disgrace.

PEDRO MOTÚ

A MORPH TO ANOTHER WORLD

Five years after seeing Fela perform in London, I decided to become a journalist. A friend at the BBC suggested that since I spoke Spanish and was interested in Africa, I should try Equatorial Guinea. English-speaking news organizations rarely sent anyone there and, he added, it was possibly the queerest little place in Africa. Often confused with nearby French-speaking Guinea, Portuguese-speaking Guinea-Bissau, and (on the other side of the planet) Papua New Guinea, this former Spanish colony was about as far off the map as I could go. It seemed like an excellent place to start a career.

Nestled in Africa's sweaty, forested western armpit, Equatorial Guinea is disjointed, split between Río Muni, a squarish coastal section of the African mainland between Gabon and Cameroon, and some islands in a line of volcanoes that runs from Mount Cameroon southwest out into the Atlantic Ocean. After Bioko Island, where the capital Malabo sits, comes the twin-island state of São Tomé e Principe, then tiny Annobón Island, which is again Equatorial Guinea. These all lie in the oil-rich Gulf of Guinea, which mapmakers consider to be the center of the world: where the prime meridian running through the Greenwich observatory in London meets the equator. This gulf is now one of the world's most exciting oil exploration frontiers.

To fly to Malabo, I first stopped in Douala in Cameroon, where I got talking to a taxi driver. He was Equatorial Guinean and had fled his repressive home country. He turned around to face me: "You, journalist! You must help us! You must tell the world what is happening in my country!" He was almost shrieking.

I made the short hop from Douala to Malabo. From my aircraft, which dodged large anvil-shaped rain clouds that mass over the Gulf of Guinea, I could see my destination, Malabo, which looks like Toy Town, nestling at the foot of the brooding, forested volcano that forms Bioko Island. The town center is a modest lattice of parallel colonial-era streets on a cliff above the sea, surrounded by disorganized slums. You can see a fair bit of what is going on by standing on an old bench near the post office and looking around. To the right, as you look out across the small bay in front of Malabo, is African Unity Point, near the Presidential Area, where it is dangerous to trespass. To the left are hotels and shops, and some of the town's nicer suburbs.

Britain founded an antislavery station here in the nineteenth century and called it Clarence, before ceding the malarial graveyard to the Spanish, who renamed it Santa Isabel. The buildings, mostly three or four stories high, reflect this Spanish influence: grubby but gracefully aging colonial architecture with bell towers and pretty wrought ironwork. Pidgin English is spoken here, along with Spanish, Bubi (from Bioko) and Fang (from the mainland), and a bit of Igbo from visiting Nigerians. There is a part-Creole population of "Fernandinos," named after the Portuguese explorer Fernão do Po who landed here in 1471 and by whose name Bioko was once known: Fernando Po.

Forest covers most of the island, and it is spectacular: 60-foot-high bamboo thickets, giant pendulous breadfruit, and trees with trunks as thick as freight trains; in places the whole lot is covered by a thick coat of creepers, like green icing. Bats and giant butterflies surf the warm, damp air, and under the canopy the warm forest is a scuttling world of bright caterpillars and snakes, redheaded lizards, pied crows, and spiders whose webs push your face back if you walk into them. In the forest I saw a dead crow, hanging upside down with its eyes gouged out, and magic charms and puppets on stakes, which people avoided. Out of town, on the mountain roads that curl around the flanks of Bioko's main volcano, forest vendors sell bush rats ("ground beef") and edible snails as big as ice cream cones, which they hang in clusters from sticks. While traveling between towns in a public taxi on the mainland, a fellow passenger stopped to buy a two-foot shank of dark meat hanging in a little grass shelter by the road: it was a gorilla's leg. The passenger, who said that he would feed it to his family, rested the bloody end against my rucksack. It felt harder than a living man's leg, but its hairy skin was still supple.

As I explored, I stumbled upon quirk after quirk. My room at the Ureca Hotel had muddy footprints on the ceiling. The phone book listed its few hundred subscribers in order of their first names, with its biggest section under "J," to fit the Josés, Javiers, and Juans. Several mixed-up names—Florentina Kuntz, Fernando Jones, and Emílio Barleycorn—hinted at the convoluted colonial history. Books about the place had discouraging titles: *Equatorial Guinea: An African Tragedy,* or *Tropical Gangsters.* A wonderfully evocative piece in the *Boston Globe,* published weeks before my visit, had the headline, "A Land Shrouded in Horror."

Westerners here were as jaded as any expatriates that I had met. "There is not even a sense of law—or anything to say, 'this is how to run a business,'" a Scotsman told me, describing his struggles to deal with the tight group of relatives that surround President Obiang Nguema. "It does not matter what business you want to do—you have to meet the same 10 or 12 people." A diplomat, already deep into his whiskies by early evening, grumbled a sweeping verdict. "Equatorial Guinea," he slurred, "is a festering pustule in the armpit of Africa."

Malabo's streets were almost empty of cars, and it was destitute and pervaded by quiet fear. The information ministry assigned me a minder who wore flip-flops and followed me doggedly to interviews. In the streets, scabby dogs snuffled in sewage while a few West African traders hawked cheap watches and statues, and scowling men in mirror sunglasses prowled in the hot stink. I saw visible signs of torture; one man could only shuffle slowly because of some harm they had done to his legs and feet. A nervous old man I interviewed would not speak President Obiang's name but instead drew an X in the air with his fingers. Near the Presidential Area, a man told me to hand over my film after I had photographed a cathedral in central Malabo; afterward my minder's hands were shaking, and he warned me that this man could "make your eye pop out." The name he whispered, Armengol, turned out to be President Obiang's brother.

Yet it still felt safe for me: police states like this have low crime rates because the criminals are afraid, too. I was able to walk unmolested after midnight through Malabo's silent backstreets, listening to my cassette Walkman. But it was quite eerie.

The Berlin Wall had fallen less than four years earlier, and it was a time of sweeping political change in Africa, as old certainties provided by three rival power systems were shifting. The first of the three, promoted by the

Soviet Union, gained prominence in large parts of the continent primarily because its revolutionary nature was useful to anticolonial struggles. But it was doubly unsuited to Africa: first, as a way of organizing the world that failed to take greed and human nature into account, and second, as a concept forged in Europe and put into practice in the Soviet Union, and thus an alien import. The second was a French scheme of awesome cunning, sophistication, and reach: a secret and tentacular creation, fed with African oil, that helped France to keep an iron grip on its former African colonies after independence and to keep Soviet expansionism, as well as British and American encroachment, firmly in check. Few people in the English-speaking world today are aware that this giant Francophone underworld ever existed: it was, in a sense, the biggest secret in Africa, and I would not step through that particular looking glass for four more years. Spanish-speaking Equatorial Guinea, however, was just outside the francophone orbit so its influence was weak; the French had as much trouble trying to get anything done here as anyone else did.

A third way of organizing the world is what most people in the West are familiar with: open markets, economic and political freedom, and the rule of law. This had clearly not rooted deeply in Malabo, either.

Over the centuries, competing foreign powers had often tried to pull Equatorial Guinea into their orbits. At one point during colonial rule the Spanish here exported up to 40,000 tons per year of the world's finest cocoa, grown on Bioko's fertile volcanic soils. The colonizers grew rich, but scores died from malaria and yellow fever; and the lives of the conquered inhabitants were far worse.

The residue from this colonial wrestling, and years of dictatorship, had—as my BBC friend had warned me—turned this into a peculiar place. Yet it was after I left that I was to have my oddest experience of all. Back in Cameroon, a man in the Catholic guesthouse where I was staying heard me speaking Spanish on the telephone as I was finishing research for a story about Equatorial Guinea for the news agency Reuters. We struck up a conversation. Pedro Motú was an ordinary-looking fellow with slightly receding hair, a denim jacket and blue jeans, and he was about to return home after a long exile. He wanted to know more about what the atmosphere was like back in Malabo, so we went out for a meal to chat.

After taking power in 1979, President Obiang had persecuted Pedro, and he spent a long time in prison—sometimes in a hot cell too short to lie down in, where he had to relieve himself on the floor. He fled Equatorial Guinea, but after the Soviet Union disintegrated, as South Africa's whites prepared to exchange apartheid for multiracial democracy, and as other Africans began to throw out old dinosaurs, Pedro took heart. Even Obiang had been sniffing the political winds and had introduced multiparty politics in 1991. New parties were setting up in Malabo, with tumbledown offices, kitschy flags, and shabby, earnest cadres with verbose and eccentric prescriptions for a better future. Pedro belonged to the largest, the Union Popular, whose main rival was the Partido Popular (its diminutive, egotistical leader Severo Moto would, many years later, be at the center of two failed coup plots involving foreign mercenaries.)

The activists got coaching from the American ambassador John Bennett, who took them under his wing and wanted to make the most of what was considered one of the world's worst postings. (Once, in 1971, a U.S. diplomat had murdered his gay lover, a fellow envoy, with scissors at the embassy.) Bennett could hear screaming from the prisons, sometimes all night; he showed me photographs of activists with feet swollen from beatings with rubber truncheons. At the time, there were 30 or 40 Americans on the islands, mostly missionaries.

"Bennett sometimes seems like a throwback to another era," the *Boston Globe* had written in a recent piece, "an intrepid diplomat struggling far from home to see human rights honored in a nation the world would just as soon ignore." The piece focused on Plácido Micó, a lawyer and a brave opposition activist. "When Míco was released from prison last year, he walked down the street here with others, a knot of shaven-headed men, finally freed from their torturers. The American ambassador, whom many have credited with helping to save Míco's life, spotted the group. 'It was an emotional thing,' he remembers. 'Are you Plácido?' the ambassador asked. 'Yes,' Plácido said. 'And are you Bennett, the American ambassador?'"

Bennett enraged the regime. "You will go back to America as a corpse," said a note thrown at him from a car. Locals in Malabo were mostly grateful to him, though some were less kind: a Spanish official, chuckling at the absurdity of trying to get a dictator like Obiang to cede power willingly, called Bennett "Quijotado," after the fictional harebrained Spanish knight Don Quixote, who tilted at windmills.

One day in 1993, near election time, Bennett went up to Malabo's old cemetery under the hill. Britain, which had no embassy there, had asked him to tend the graves of Flight Sergeant J. J. Duffy and crew members of a British Royal Air Force Sunderland flying boat that had crashed nearby in 1944. On this visit Bennett noticed Obiang's brother Armengol and some aides flitting behind gravestones, watching him.[1] Armengol went back and publicly accused Bennett of witchcraft, and the interior security minister said that Bennett had been caught taking traditional medicine in the cemetery to influence the election. It was a ridiculous allegation, but Bennett did not last long: the embassy was soon closed in a budget-pruning exercise. As it happened, the thorn in the relationship was removed just as a large American organization was moving in: Mobil Oil.

Pedro Motú was nervous about going back home, and as we ate chicken and chips in the Douala bistro, he asked me what I had seen of the emerging multiparty scene. I reassured him that while there had been some torture and beatings, plenty of activists were operating openly. We drank some beer, and returned to the guesthouse, where I recorded an interview. Less than a week later, on July 25, 1993, Pedro flew back to Malabo.

He called in on Bennett. Pedro was, he said:

> in excellent shape, had shaved his head, and spoke with real passion about the need to change the dictatorship to a democracy. I am not sure he was quite a democrat, but he sure projected charisma; I was impressed as we drank tea together in my office. He said the Union Popular [his party] was already backed by the majority in the country. . . . Of all the "pols" I met in Malabo during those three years he was the most militant in outward expression—something which I am sure got back to Obiang from other sources. While he spoke of politics and party and elections and so on, he seemed prepared to go beyond that, to physical struggle, if necessary. I was fascinated by him. He was ready to risk all, and openly confrontational.

A month after Pedro returned home, he was dead. The government said that he had killed himself in prison, which I found hard to believe. And a strange story began to leak out. It is bizarre, indelicate, and it even sounds insane, and regrettably it fits some of the worst clichés about Africa. Yet this is what I have discovered.

One afternoon in late August 1993, Pedro visited the Ureca Hotel where I had stayed, to visit his party leader in a hotel room. Suddenly, one of Obiang's aides turned up with soldiers.[2] They broke a hotel door down, caught Pedro, then dragged him down the stairs, his head bouncing down the marble. Guests saw him being beaten with rifles and dragged to a police car, with blood flowing from his nose and ears. "I'm dying, I'm dying," he said as they bundled him in.

Bennett remembers:

Some of the Union Party followers came to the embassy seeking me out in hope that I would intervene, which I would have done had I been in town. But no one from anywhere raised a ruckus as we had successfully done when Plácido Micó was arrested. After I returned to Malabo, a young man came by the embassy to see me. A political activist, he told me that he had been in the same common room—sort of a small hall apparently—where Motú was dumped on arrival. My informant—who seemed genuine and claimed that he was curled up in a shadowed corner, left there from an earlier torture session (petrified that he would be noticed)—said that the police . . . came into the room several times and beat Motú without mercy with cudgels or staves. Apparently, his tormentors revived Motú with buckets of water when he went unconscious and the beating continued. My informant said that the attackers were yelling at Motú, who did not respond beyond grunts of pain when hit. The beating sessions were nothing like interrogations, but rather were personalized rants against him. Motú died sometime during the night.

According to testimonies told to the United Nations special rapporteur on human rights, and published in a U.N. report, four doctors then removed Pedro's brain, heart, and genitals and they were used for "ritual purposes."[3] His body was never returned to his family for proper burial, a grave offense to local traditions. Other sources go further, saying that Pedro's body parts were then eaten by politicians. The government said that Pedro had committed suicide when "confronted with the enormity of his crimes." Pedro had a late-teenage son who stayed in Malabo after that, often on the run from the police.

I cannot prove that this story of modern cannibalism is true. But I do not see any particular reason to doubt what several people have told me, backed up by the United Nations report.

How could this happen in the modern age? The reasons, locals tell me, lie in history. When independence came in 1968, Equatorial Guinea fell under the spell of a poorly educated man, Macias Nguema, a violent, stick-thin despot who descended into paranoid madness and called himself the Great Sorcerer, whom nobody could touch. He said he would walk into fields at night to talk to ghosts; he burned books "to weed out the devil [Spanish] culture," closed churches, arrested priests, and confiscated seagoing transport, including fishing canoes. An African Caligula, Macias proclaimed himself God's "Unique Miracle" and killed 40,000 people—a tenth of the population—including 21 of his cabinet ministers. He had people crucified on the airport road to show visiting diplomats; being a journalist was a capital offense. Foreigners were ransomed; he reportedly got $57,000 for a German woman, $40,000 for a Spanish professor, and nearly $7,000 for a dead Russian.[4] At Christmas in 1975, he had 150 prisoners executed in the stadium, as a band played "Those Were the Days." Toward the end of his rule, Amnesty International called Equatorial Guinea "an immense field of torture, from which the only exit is the cemetery."[5]

Even then, foreign powers jockeyed for control. Russia and China sent advisors, and a friendship grew with North Korea, a kindred pariah. Spain, while embarrassed by Macias, still battled to keep a foothold in its only former sub-Saharan African colony. French interests lent Macias money in exchange for a big forestry concession,[6] and supplied arms and vehicles for the police.[7] Equatorial Guinea entered the Franc Zone, a club of African countries whose currency was fixed against the French franc and that surrendered monetary control to the French central bank. Telecommunications came under French influence, and a French cultural center in Malabo brought art exhibitions and prancing troubadours. Relations soured with Spain, and the newspaper *El Pais* called the government "a group of individuals whose only ideology is institutionalized pillage." Spanish diplomatic pouches were opened and its officials routinely slighted; some of these humiliations, like the unpunished rape of a Spanish aid worker, were more than symbolic. While Britain's *Sunday Times* was calling Equatorial Guinea "the wickedest place on earth," French media was exultant over its "capture."[8] Nevertheless, French influence never penetrated too far into this strange, vicious little nation.

In 1972—according to British newspapers and a subsequent interview—the author Frederick Forsyth had advance knowledge of plans to topple Macias in a mercenary-led coup.[9] While reporting on the Biafran war in eastern Nigeria, Forsyth had befriended the Biafran secessionist leader Emeka Ojukwu, who was defeated by Nigerian federal forces in 1970. The bizarre plan was to replace Macias with the Nigerian general. The plot fell apart, but Forsyth used his knowledge to write *The Dogs of War,* whose fictional town of Clarence was a thinly disguised Malabo, and whose president, Jean Kimba, "mad as a hatter, and nasty as a rattlesnake," was Macias.

By 1979, a third of the population was dead or in exile, and Macias's nephew and right-hand man, Obiang Nguema, mounted a coup. Macias bolted into the jungle, but he was tracked down and captured by an army unit, led by my friend Pedro Motú.[10] At a show trial in the local cinema, Macias cursed those in the courtroom and vowed to return to haunt Obiang.

(A Reuters veteran said that most of the reporters who covered the trial came down with malaria.) Macias was executed, and Obiang took over as president. Obiang has since called himself "Liberator of the People."

After that, Pedro was considered a threat. "I think Obiang saw in Motú not only a political opponent, but—perhaps more important—he was highly jealous of him," Bennett said. "Motú had proved himself competent in combat—there was some fighting as his force made its way from Bata to Mongomo, including possibly with a Cuban detachment—and he was popular as a result of his deed. Of course, Obiang could not permit other stars in the firmament."

But Pedro had a more important significance: it was widely believed that because he had been able to capture Macias, whom Obiang had called "an envoy of the devil and president of sorcerers," Pedro must have possessed special powers.[11] An opposition official said that he thought Pedro was planning to deliver Macias's amulets to his party leader on that fateful day. The spirit world is alive in Equatorial Guinea; bodies and organs can transmit inner forces, and consuming Pedro's parts would pass on these powers. As several Equatorial Guineans have told me, such powers were considered vital at a time of such wrenching change.

Over millions of years, rivers on the African mainland deposited organic sediments into Equatorial Guinea's offshore zone, sediments that, pickled under tremendous pressure and heat, slowly became crude oil. From the late 1950s plenty of oil had been found in neighboring Nigeria, Gabon, and Cameroon, and though firms from Spain, France, Britain, Nigeria, and Chile had found little, it seemed likely that oil lay further out, under waters that were too deep for drilling rigs to reach.

When I first visited in 1993, this was changing: deepwater drilling technology had advanced, bringing new maritime areas into range. A couple of years before Pedro's death, Walter International, an American wildcatter, had tapped a small field of valuable light oil near Bioko and was exporting occasional cargoes. Ambassador Bennett discovered that Walter International was paying for Obiang's son to study at Pepperdine University in California, and the company was grousing that the lad was "spending at will."[12]

Unexplored backwaters like Equatorial Guinea stir great hopes for oil companies. It is hard for inexperienced African countries to know what a good deal is. Because they tend not to have clever lawyers and geologists, and oil contracts are by their nature secret, they have nothing to compare them to.

"By 1994 there were some major companies interested in coming in," Bennett said. "Statoil came but decided not to invest because of the corruption. A very experienced industry guy came in; he had worked for several of the oil majors, and said, 'This will be a gold mine!' I told him about the politics. They had no interest in the politics."

Two years after Pedro's death, a Mobil (now ExxonMobil) rig drilling north of Malabo near the maritime border with Nigeria struck Zafiro, a billion-barrel field of sweet, light oil ideal for making gasoline in American refineries. The companies had obtained fabulous contract terms: the World Bank described the Mobil contract (which it had acquired in 1994 from another American company, United Meridian Corp.) as "excessively generous to the operator."[13] Drawn up in secret, without open tender,[14] whatever else it was, it did not appear to be a shining example of how free markets should work.

Obiang dithered about confronting Mobil, and first sought to squeeze out extra cash through other means. Some came directly from the oil companies, who would first lend Equatorial Guinea money (Mobil called it "bringing forward cash payments")[15] in exchange for being allowed to modify their future contractual obligations—perhaps as tax deductions, or with recalculated repayment terms—to get the money back later.[16] The companies were acting like giant banks, lending money at implicit interest rates of up to 19 percent, and possibly more, racking up huge extra profits.[17] Relentless demands from the rulers for ready cash meant that these deals kept rolling in.[18]

The wealth flowed in other ways. Obiang's brother Armengol set up a company to provide security guards, and the oil firms paid for his services. Relations of the ruling family set up employment agencies,[19] which ensured that oil companies did not employ "enemies" in the oil sector, and which harvested large percentages of the oil company workers' wages as "fees." Periodic government decrees to multiply the minimum legal wage boosted these agencies' profits. The oil companies hated these bloodsuckers, which made their workers more expensive and unhappier, but they accepted them. Obiang also had a cement import and construction monopoly,[20] which cashed in on the building boom that followed the oil boom, and he owned hotels, restaurants, and shares in the oil business itself, in commercial partnerships with Mobil.[21] Members of Obiang's inner circle teased out cash through unofficial import duties, port charges, monopolistic import and sales schemes, and hotel ownership (one company bulldozed a school to make way for a hotel), and some used prison labor to keep costs down. As the cash began to build up, foreign bankers began to invite them to stash their new wealth offshore.

The second big change in the flow of cash came later, in 1998, after Obiang, fortified and feeling more confident with his new-found riches, ordered the Mobil contract renegotiated.[22] He was understandably piqued that Equatorial Guinea had only received a small percentage of the value of its oil as revenue the previous year.[23] IMF figures—which may not tell the whole story—put Equatorial Guinea's share at 13 percent, which is staggeringly low by regional standards.[24] Wood Mackenzie, a respected British oil consultancy, published some details of the new deal.[25] Based on estimates of 350 million barrels of oil, at $16 a barrel, Wood Mackenzie reckoned that the renegotiation would eventually bring Equatorial Guinea an extra $650 million or so. Since then, however, Zafiro's recoverable reserves have been revealed to hold more than a billion barrels, making it (by one measure) the largest oilfield in sub-Saharan Africa, and the oil price has quadrupled.[26] So—making some major but not unreasonable assumptions (which cannot be avoided since the contracts are secret)—this renegotiation *alone* might bring Equatorial Guinea $5 billion in extra money over time.[27]

When the first contract was signed in 1992 (not, remember, by Mobil but by a predecessor), this much money was worth perhaps 50 times Equatorial Guinea's gross domestic product, or well over a lifetime's output for every man, woman, and child in the country. One might compare this to the colonial treaties of old, where European empire-makers won huge territories in exchange for cloth, axes, and cutlasses. This time the outsiders were Americans, who did not use colonial soldiers, but lawyers, geologists, and accountants.

A Mobil official said the 1998 deal represented "a satisfactory outcome for all concerned" and that it "helps address the government's concern for the country to see tangible benefits from oil production at this early stage."[28] Yet even after the renegotiation, the new Zafiro contract was *still* hugely favorable for Mobil[29]: the following year the IMF said that the government's share remained "very small by international standards. . . . The new contract continues to bestow the oil companies with, by far, the most generous tax and profit-sharing provisions in the region . . . and [the contract's] management continues to lack transparency, with no fiscal control over the payments made by the oil companies." For whatever reason, Equatorial Guinea repeatedly rejected offers of technical help from the World

Bank, which could have helped with serious analytical firepower to challenge the contract.[30]

As Zafiro's oil sailed across the Atlantic to American refineries, then into American cars, another armada of superprofits followed.

ExxonMobil officials say that governments in virgin exploration zones must offer special terms to offset the oil company's risk of spending money on exploration and possibly not finding oil. This, they argue, is why contracts like the original 1992 contract that Mobil acquired might seem unfair. "The greatest incentive for doing business in any part of the world is the assurance of expected revenues in the event of success, bearing in mind that our successful ventures must also pay for unsuccessful investments," Ken Evans, ExxonMobil's vice president for sub-Saharan Africa, told governments in the region. [31] Yet it is hard to see a fair balance of risk and reward here. If the companies had found nothing, they might have been $30–50 million out of pocket in exploration costs. But Zafiro's billion barrels of oil—even at a conservative $30 to $50 a barrel—is worth one thousand times more. And, with big discoveries nearby (Zafiro falls in an area that oilmen call the "Golden Rectangle"), the companies were quite confident of striking it lucky.

As Equatorial Guinea took its faltering first steps into the oil age, I wondered if oil would make Obiang's brutal government behave better. Locals were wondering the same thing: one activist in 1993 told me that he hoped oil would attract American attention and that once America saw what was happening, it would invade to overthrow Obiang. Others had more realistic dreams. Perhaps, some said, oil would modernize this closed nation and generate a trickle down of wealth, giving ordinary people the financial and political independence to help them prize their rulers' fingers from their throats. Or, others thought, even modest oil revenues would go a long way in such a small country. "Even with corruption," a French diplomat said, "so much money will come that things will happen here." Maybe, some people thought, a combination of easy wealth and international attention and exposure would soften the dictator.

Others had faith in free markets. Most investors shun the worst-ruled countries; they prefer low interest rates, good infrastructure, trustworthy contracts, and a benign economic and political environment. These incentives for rulers to shape up often work well, and many argue that they are a key to prosperity around the world. On my first visit to Equatorial Guinea this vision was widely shared in the West; it was an era of great faith in the transforming power of private enterprise. Oil companies—especially American ones—love to hitch themselves to this positive vision, arguing that their investment brings positive engagement and drags backward countries into the modern world. In the popular mythology, the extraction of hydrocarbons is the quintessentially capitalist occupation; it evokes images of bold, risk-taking Texan wildcatters and the finest traditions of entrepreneurialism; the product itself, this legend holds, oils the wheels of free enterprise and makes the world richer.

I knew very little about oil in those days, and there was not much academic research on the subject. I suppose I entertained similar hopes for Equatorial Guinea.

It would take me some time to unmask the deception. Oil companies do not base their investment decisions on whether a country is well run. They go where the oil is. If the business environment is bad, they just demand better contract terms to compensate for the extra risk, cost, and trouble. Profits from oil are different from normal industrial profits, since they tend to far exceed the costs of getting the stuff out of the ground. Economists call these superprofits "rents"—a bit like free money that you don't have to work for. The Polish writer Ryszard Kapuscinski put the problem like this:

> Oil kindles extraordinary emotions and hopes, since oil is above all a great temptation. It is the temptation of ease, wealth, strength, fortune, power. It is a filthy, foul-smelling liquid that squirts obligingly into the air and falls back to earth as a rustling shower of money. Oil creates the illusion of a completely changed life, life without work, life for free. Oil is a resource that anaesthetizes thought, blurs vision, corrupts. People from poor countries go around thinking: God, if only we had oil! The concept of oil expresses perfectly the eternal human dream of wealth achieved through lucky accident, through a kiss of fortune and not by sweat, anguish, hard work. In this sense oil is a fairy tale and, like every fairy tale, it is a bit of a lie. It does not replace thinking or wisdom.

Although I don't agree with some of what is in Kapuscinski's beautifully written books,[32] he is dead right about oil. Low interest rates and inflation, or other normal parameters of responsible economic management that attract the oil companies or determine the level of reward, don't matter much, so long as the oil is there. To get tax revenues, rulers no longer need to worry about improving the local business environment—and, by extension, the lives of their citizens—a point that Fela

Kuti had instinctively understood in the 1970s. Oil turns the benign incentives offered by other kinds of industries on their heads. It is among the fiercest anticapitalist, antidemocratic forces in Africa.

There is another thing about oil. *Consumers* of oil benefit by being able to burn it in their cars and boilers; these benefits are dispersed widely through society. But for *producers* of oil, the benefit is money, which comes from a point source, controlled by the rulers, who decide how to share this out as they see fit. The benefits of Africa's oil are not bottom-up, but top-down. (Oil might serve as a useful control experiment for foreign aid, which can suffer from many of these problems too.)

In Equatorial Guinea, oil also began to generate strange eddies in the power structures. In such places, modern forms of government are just façades for the real, hidden power flows. French academics talk of a "rhizome state," where ministries and government departments are just the visible parts of much larger subterranean systems, whose strength lie in their tangled roots. Decisions here are often made after Obiang consults clan elders in Mongomo; just before my visit in 1993 one of the biggest fights was between Obiang's family and that of his wife Constancia, over where the ancestral skulls should be kept. The archbishop helped broker a settlement—not between Obiang and his wife, but between the tectonic plates of the Mongomo clan.

Equatorial Guinea was poised between two worlds, utterly alien to each other. When Bennett reported back to Washington, some surely saw him as an eccentric; perhaps slightly mad. "Obiang seems to be working in a dimension where angels fear to tread," he told me over coffee back in Washington, years later, reminiscing with a smile. "It was a morph to another world."

What might result from this uncomfortable combination? In the years since 1993, I have watched these two worlds collide, with results even more bizarre and shocking than I could have imagined.

3

ABEL ABRAÃO

WIELDING THE OIL WEAPON

In Malabo I had borrowed a typewriter and sent some news reports to Reuters, using the post office fax machine. They invited me in for an interview and offered me a job as their correspondent in Angola, an oil-rich former Portuguese colony five hundred miles south of Equatorial Guinea.

Portuguese-speaking Africa had long held a vague sense of menace for me; refugees from the bush war in Mozambique came into Malawi where I lived as a child, and I sometimes heard stories of heavy fighting in Angola in BBC radio broadcasts that my parents listened to in the dining room. I remember sitting in my grandmother's house in Britain during a school holiday, staring with boyish excitement at television pictures of an armored South African military column pushing north into Angola in support of Jonas Savimbi's rebel National Union for the Total Independence of Angola (UNITA).

It was 1993 now, and UNITA forces were rampaging across the country after the collapse of a peace agreement. The United Nations was calling it the world's worst war. Becoming the Reuters man in Luanda was an intimidating prospect, but I have also long been attracted to adventure, and there was nothing to beat this one. I did not yet realize that the most fulfilling adventure would not be the obvious one—flying into war zones—but the quest for knowledge: fathoming how oil-rich countries like this function. This was, and still is, intellectual territory that is vastly underexplored by outsiders. After three months' intensive journalism training under the legendary late George Short at Reuters headquarters in London, I flew in to the capital, Luanda.

UNITA had severed the inland roads, and Luanda was a self-contained coastal bubble, housing the seat of power and four million people in a town originally built for under half a million. It took time getting used to being in an African country where many people, after five hundred years of colonial domination, do not speak an African language; urban Angolans often speak only Portuguese, and many are light skinned. In southern Angola I saw white Angolan peasants digging in fields with hoes: they were the Angolanized descendents of 350,000 Portuguese colonists who had built up the world's fourth largest coffee export industry and a thriving manufacturing sector before independence in 1975. Angola was also, according to recent academic research, the origin of the "20 and odd Negroes" who in 1619 became the first African slaves to set foot in North America.[1] Up on a hill above the sea, near the British embassy, you can still visit the old slave fortress in Luanda, through which many tens of thousands of slaves passed over the centuries.

In the city center, elegant arched residences and decorative mosaics intermingle with more workmanlike six-story tower blocks; and unkempt vertical villages, festooned with washing lines, radiate outward and upward in a moldering jumble from the Baixa, a central zone near sea level. A seafront road arcs around from the port, past the beautiful pink-domed central bank, then onto a spit of sand called the Ilha, where wealthy Angolans, aid workers, and oil expatriates spent their weekends. Outside the town center were the *musseques* or slums: Dickensian forests of corrugated iron, cinder blocks, and sewage.

The cost of living staggered me: one hotel offered me scrambled eggs for $32, and a plate of medium-size prawns cost me $1 per prawn. Expatriates paid $1,500 a month for a sweltering apartment yet still had to use home generators and tiptoe across stinking, refuse-filled puddles and up slimy unlit staircases to get home. This high cost of living was an example of the oil-fed Dutch disease that had killed agriculture in the 1970s oil boom in Fela's Nigeria. (Things have gotten worse: nowadays expatriates can pay $10,000 to $15,000 *a month* for pleasant but unremarkable houses in decent areas.)

Almost all of Angola's half-million barrels per day of oil production was offshore, and the war hardly touched it. The world's oil companies were then sniffing at a large area about 70 miles offshore, where the water was over a mile deep and where seismic studies had revealed "flat spots" in the geology, which indicate an interface between oil and water, or between water and gas. Drilling for oil is never precise; you can identify geological caps that trap oil as it slowly seeps upward through ground, but even with the best seismic information you are never sure if there is oil before you drill. Flat spots are as good an indication as you are going to get that oil or gas is down there, and in Angola's case the seismic data had the companies slavering. With no hurricane season, waters that were filled with giant jellyfish and hammerhead sharks but were otherwise untroubled, this was soon going to be the world's premier stretch of oil exploration real estate.

Luanda was awash with weapons—when the 1992 peace agreement collapsed, the army unloaded up to a million guns into the loyal slums—but the town was surprisingly safe for me; I often walked several city blocks, alone at night, to get a plate of gristly beef, fried egg, and warm Sagres beer. Nevertheless, violence was close to the surface. Outside my hotel window they caught a thief, and I remember a bystander taking off one of his shoes and then reaching into the scrum with it to bash the miscreant on his profusely bleeding head. What most disturbed me about it was the sight of a small girl in a frilly dress, holding her father's hand and jumping up and down with excitement at the spectacle. The greatest menace for me was the police: they shot at cars that did not stop at roadblocks, and they tried outrageous extortion on those that did stop. One evening four armed policemen opened my car doors, got in, and ordered me to drive. It was dusk, and rumors had been circulating of car hijackers posing as police. Fortunately, we passed my hotel, whose owner had recently hired armed security guards during a commercial dispute. I had been tipping these fearsome rogues, as a vague insurance policy, and when I stopped and called for their help they came, grinning and fingering their Uzi submachine guns. My "police" passengers jumped out and sprinted off. I learned how to tackle the roadblocks: always smile, talk calmly, have cigarettes to offer, and maneuver the conversation toward women or football. Behind the intimidating sunglasses lay a typical Angolan: desperately poor but ultimately warm, and looking to feed his family.

The city throbbed with people. Refugees and orphans crammed into every free dark hole, including the sewers. Once, I saw a scramble for football tickets so intense that a man crawled right over the heads of people in the sweaty scrum to get to the front. From the Reuters car window street life appeared on three levels: adults above head height, hawking things like car antennas or toothpaste; street children begging at window level; and then, level with the dogs, were polio victims and *mutilados,* the living scars of war—men who had trod on landmines and had been kicked out of the army, and who were now reduced to scraping along the ground, pestering motorists for spare change. Driving with the windows down was a voyage through a patchwork of smells: chemical fumes might be followed by freshly-baked bread, then a mixture of rotten fruit and excrement, which would hit me in the face like a bad washcloth.

Scabby children scampered between potholes in bare feet and diesel-covered jackets; you saw their upside-down legs wriggling from the tops of rubbish bins as they ferreted for unmentionable treasures among the rats and warm filth. These children surrounded me in alleys murmuring *Amigo, Amigo,* gently tickling my elbows and pleading for cash; if my resolve broke and I handed out kwanzas I would never know peace again on that street corner. Sometimes it was not *amigo* but *chefe,* or *patrão,* elevating me to the rank of boss (thus, they reasoned, obliging me to fulfill my cultural role and give them their due as underlings). I learned to stick with a few regulars, hand out kwanzas from time to time, and make the rules clear. Angolans jostled every day to pull off *esquemas*—schemes (or perhaps scams) in desperate daily struggles to get money.

How does money flow down through an economy like this? One way to visualize it is to think of islands in a big river of money, with people on their own islands trying to capture some of the flow to irrigate their own crops, in their own little *esquemas.* Or you can envisage a giant network of pipes shaped like a Christmas tree, with life-giving liquid flowing in through a big pipe at the top and being subdivided at different nodes. The pipes get smaller and more numerous on the way down, and people guzzle at their open ends. The fat cats perch at the top, sucking from the fattest pipes; at the bottom thousands jostle for access to the tiniest ones, while millions crowd around them hoping to get something. Each node has a tap, controlled by a gatekeeper, often a bureaucrat, who drizzles the liquid down to underlings, in exchange for their loyalty. This is a political structure: the relationships are between rulers and subjects; wealth is obtained from political power (one's position in the tree,) and vice versa.

The Angolan tree has many vertical connections but few horizontal ones; cut one link—sack a minister, for example, and a whole chain of people dependent on that link falls down. The new minister typically sweeps away his or her predecessor's supporters, and appoints new, loyal ones. In a rich western country, by contrast, each node in the tree has many horizontal links, and this gives the structure its strength: cut a link, and the structure holds together. People don't need to be obedient to flourish, and the key relationships are not between rulers and subjects, but between fellow citizens. The government's role is to oversee and enforce the subtle rules and conventions that nurture the horizontal links, allowing private operators to interact and generate production, growth, and employment. These horizontal links generate the trust between citizens that was so badly eroded by the oil boom in Fela Kuti's Nigeria.

While Nigeria fragmented between ethnic and religious groups, none of which were dominant, Angola was a bit different: power was centralized under President José Eduardo dos Santos, a Soviet-trained engineer who took office in 1979 and only formally abolished Marxism-Leninism in 1990, three years before I arrived. In Luanda the crucial fragmentation was between two budgets: the money controlled by the president, and the remainder, which gurgled chaotically down through the ministries, into Angolan society, and offshore. From my street-level vantage point I knew little of what went on at the top of the economic Christmas tree: the headwaters of the oil money. These were financial operations of tantalizing mystery, which I would only peek into years later.

Many of the secrets lay in a low-rise beige building in the Baixa. This was the headquarters of Sonangol, the octopus-like state oil company that oversaw 98 percent of exports (most of the rest was diamonds),[2] and whose oily tentacles reached around the globe, tugging discreetly at levers of power in the capitals of the West. If *Futungo*, the presidency, was the brain of the system, Sonangol was its heart, pumping cash around the different arms of the economy. Sonangol was also a black box—the basis for a vast, invisible budget that bypassed parliament and the treasury, and flowed instead through offshore structures at the behest of the president. He used it to buy arms, to pay for the intelligence services, and to carry out secret financial operations to bolster his power. In this building I struggled for two years with a surly, evasive bureaucrat who set humiliating obstacle courses in the way of my getting even the most basic oil data for Reuters. This was a system steeped in old Soviet suspicion: in a time of war, strategic information on oil matters was a matter of national security. Executives from the western oil companies scuttled away when I approached.

An oil official described how a small part of the money got diverted back in those days. With American oil companies like Texaco or Chevron, which then produced most of Angola's oil, negotiations for oil contracts took ages; once they were nailed down, the nuts and bolts of oil operations ran smoothly enough. Negotiations with the French company Elf, by contrast, were fast, but they left loopholes and vague terms, opening up spaces for informal negotiations. One such arena, which will be familiar to people who fill in tax returns, is this multibillion-dollar question: What counts as a cost of production, which can be recouped as expenses? Drilling rigs and helicopters? Certainly. Nice cars for oil executives in Luanda? Probably. But champagne and overpriced corporate entertainment? (And the "social programs" that oil companies love to plaster across their glossy brochures—schools for orphans, solar panels in villages, and the like—are often deducted as costs, meaning that it is the government, not the company, that ultimately pays.) Loïk le Floch-Prigent, the former head of Elf, said that his company routinely took commissions from this arena to kick back to officials.[3]

Elf flooded Luanda with expatriates who fitted in well, sometimes outraging the scandal-loving Angolans with raunchy goings-on at the French Club on the Ilha. The Americans were different. Chevron housed most of its staff in a bubble: a tiny piece of America carved into the northern oil-rich enclave of Cabinda, with manicured lawns, ranch houses, a golf course, and a ferocious security fence fortified with landmines planted by the army (and which Chevron has repeatedly asked the government to remove). Exxon, which was just starting up, ran minimal operations in Luanda and housed its officials in dry, Americanized compounds far from the town center. Even guest lists for parties were organized out of Texas, a sure recipe for misunderstandings in this protocol-conscious country. A diplomat remembered that when the oil minister and vice-minister turned up at an oil party, Exxon executives surrounded the minister and left the vice-minister in a corner. They were unaware that the vice-minister was the president's trusted man in the oil industry and had all the power. "Battleship Exxon," as one official described them (on account of their being trigger-happy with aggressive lawyers when contract disputes broke out—which was not the African way of doing things) infuriated Sonangol, who later stopped returning their calls for a while, slowing down their giant projects. Exxon began to learn humility in Angola, and their oil operations began to run smoothly again.

After the inner cabal had taken their cut of the oil money, the rest flowed into the budget (which was then a badly typed document of economic fiction),[4] where new interest groups scrambled for it. The ministries were one set of interest groups, and they lobbied against each other to boost their allocations. Civil servants were another interest group; the inefficient state-owned companies constituted yet another. Each was plugged into, and scrambled for, the flow of oil money.

Underlings tapped into the money as it cascaded on downward. A port official needed bribes to let a container of imported cars (bought with oil money) through. The ubiquitous police roadblocks were milking points for getting the cash out of the pockets of Luanda's hapless motorists. In front of the press center a teenager named Kwanza (like the national currency that he was so good at wheedling from passing foreigners) led five rascals who begged and cleaned cars for a living and defended this lucrative, foreigner-rich turf—one island carved out in the flow of money—diligently against all comers. They could easily beg and earn hundreds of dollars a month, more than most government officials. I let them sleep in the Reuters car, in exchange for keeping it clean and safe from thieves, and they push-started it when the starter-motor broke (I had to get the spare part flown in from Lisbon). In a few weeks, the car was stinking.

Kwanza and the boys only tolerated one other beggar on their territory: a friendly, bearded madman who had lost his mind in the war and wandered about, half-naked, yelling into the sky. (Once I spotted another madman, with pink intestines hanging out from a cut in his belly, swinging as he walked. He died soon afterward.)

The poor tried their hand at informal trade, or *candonga*. Unlike in neighboring Zaire, where President Mobutu Sese Seko had told his people to "débrouillez-vous" (fend for yourselves), free enterprise was still kind of frowned on by Angola's former Marxist-Leninist rulers, who still felt that the state should ultimately use its oil money to plan and provide for its people. Official pronouncements suggested that they were embarrassed by the informal markets' chaos and squalor, and they periodically sent whip-wielding police to chase the hawkers away. The traders soon seeped back.

I soon picked up a common obsession: the fate of the town of Kuito in central Angola, where UNITA was trying to shell and starve a besieged government garrison, and up to 50,000 town residents, into submission. UNITA had reached the town center, and the main street was a front line. If the war had an epicenter, this was it.

In Luanda, we kept track of the fighting at Kuito from the dispatches of Abel Abraão, the trapped national radio correspondent. He spoke like an orator; rolling his r's theatrically over the radio crackle and using medieval language to describe the latest "flagellation" of his "martyred" city by UNITA's "diabolical" 120-millimeter mortars that poured shells—each big enough to demolish a house—into his neighborhood. Women and babies lay wounded and screaming in the road in front of family members who could not rescue them because of UNITA snipers. Abel Abraão was Kuito's voice to the world, and he drove UNITA crazy: their commanders put a price on his head, which they shouted across the front lines with loud-hailers. The nearby city of Huambo had fallen to UNITA following 55 days of fighting, after which thousands of defeated residents and soldiers walked two hundred miles to the coast in a giant, snaking armed column, to the safety of Benguela. On the trail exhausted women gave birth, then stood up and marched on. Kuito, we all assumed, was next.

The war seemed to have lost all meaning. Corruption had exploded after the end of Marxism-Leninism, sapping the army's fighting ability. Rich generals, high on arms kickbacks, would send out huge, lumbering ar-mored columns, which UNITA's wily fighters would rout in decisive am-bushes, then harvest the expensive equipment that the soldiers left behind. The generals would order new tanks and send new press-gangs into the slums for fresh youths for the army. Younger Angolans could remember only war, which started with anticolonial struggles long before independ-ence in 1975, and continued immediately after that.

The Soviet Union and Cuba backed the ruling Popular Movement for the Liberation of Angola (MPLA), while Jonas Savimbi courted the West. Savimbi spoke English, French, Portuguese, and several African languages fluently; he was well read, witty, and charming, and, like most Africans, he understood the West far better than they understood him. The bearded rebel leader persuaded America that he was a heroic Christian democrat, and an anticommunist freedom fighter; in fact he was an old friend of Che Guevara who had organized UNITA as a peasant army along Maoist lines, ruled by fear. His soldiers carved "pockets" into their captives' sides, jeering as they forced their hands down into them, and they would beat pregnant women face down, first digging a hole for their bellies. Savimbi burned UNITA dissidents on public bonfires,[5] along with their families. Graffiti in Luanda succinctly embraced the character of the two sides: "UNITA kills, the MPLA steals." Savimbi slept with his aides' wives, to put his underlings in their place and to make them worry what he knew about them. Apartheid South Africa invaded to support him, and Ronald Reagan, who sent him guns and missiles in the 1980s, once hosted him in the White House and predicted that Savimbi would win "a victory that electrifies the world." In 1985, the now-disgraced Washington lobbyist Jack Abramoff joined Savimbi in southern Angola to host a meeting of Afghan Muja-hedin, Nicaraguan Contras, Laotian guerrillas, and what one South African newspaper called the "Oliver North American Right." The super-wily Savimbi even pulled off the trick of being funded by the CIA and by Communist China, simultaneously. Donald Steinberg, a former U.S. ambassador to Angola, called Savimbi "the most articulate, charismatic homicidal maniac I've ever met."[6]

The West's hypocritical stance generated one of the Cold War's supreme ironies: communist Cuban forces defending American facilities in Cabinda, whose oil was sold to the United States, enabling a Marxist-Leninist government to buy guns from the Soviet Union to fight U.S.-backed rebels. Elf paid UNITA off,[7] yet the Angolans still gave them oil concessions, not wanting France as an enemy. The head of Elf explained the system.

> Africa is special. These countries have tribes: the head of state does not always control them all. You talk to the central power, without forgetting tribal chiefs. For the oil man, it is about respecting their system. Everyone must be happy, to avoid trouble when you produce oil. These means are, naturally, financial. Either the head of state shares money with ethnic groups or opponents, or someone else does it.[8]

The oil prize had turned Angola into one of the Cold War's hottest battlegrounds. "In Europe, the borders were set in stone and there was no opportunity of expansion there for either side—it would have started a new world war," a Soviet Communist Party official later said. "Where could the hunting take place? To be rather crude, in those areas where there was still prey. That was the Third World."[9]

If the war had lost its meaning for most Angolans, it also had many roots. Before independence, Henry Kissinger had backed the Portuguese dictatorship and declared that "the Whites are here to stay."[10] (Just as dissent was more recently suppressed in the United States ahead of the Iraq invasion in 2003, Kissinger neutered the opinions of dissident Africa experts in the State Department who could see the importance of growing anticolonial movements.) The Portuguese, thus encouraged, tried to cling on, but after a coup in Portugal in 1974 independence became inevitable; 400,000 settlers fled the approaching communist "menace," breaking factories, spoiling cars, and bequeathing a political and economic vacuum. Portuguese advisors stayed behind in just one industry, oil, which has remained a strategic island of competence ring-fenced against the chaos elsewhere. Since birth, the Angolan state has been shaped by oil, which has leached the ministries of their most skilled personnel.

Savimbi's messianic determination to rule Angola was another driver of war, which also had an ethnic strand, pitting UNITA's mostly Ovimbundu fighters from the central highlands against an army drawn strongly from coastal, urban areas. Commercial matters helped prolong the fighting; in the gunslinging diamond zones, army generals and UNITA commanders mined side by side under local non-aggression pacts.

Oil helped crystallize yet another of the war's undercurrents. Savimbi claimed to be leading "real" Africans in the bush against corrupt, light-skinned, and oil-rich urban elites who despised them. An Angolan friend showed me a photograph he took in 1992, just after the war had re-started, of him and his buddies in an open-topped Land Rover, brandishing AK–47s and clearly in high spirits. They had been "cleaning" UNITA out of the suburbs. How did they know, I asked, who UNITA was? He rubbed his thumb and forefinger under his nose: "It was their smell." Oil gave the government the means to build expensive air bridges to supply the inland provincial capitals, many of which were, like Kuito, completely surrounded in a hostile UNITA sea. This helped the war to become, literally, a war of the towns against the countryside.

Diplomats and aid workers gossiped incessantly about Kuito. I would think about it constantly, and yearned to meet Abel Abraão. But I doubted that I ever would: it seemed unlikely that the plucky little town's starving residents could hold out much longer.

Abraão's radio reports seemed unreal as I got on with "normal" life in the Luanda bubble. The city offered contrasts of wealth and misery unlike anything I had seen in Africa. Many expatriates spent their lives inside Luanda's central "concrete city," the bubble inside a bubble, which was the area built by colonial planners and was plugged into water mains, sewers, and (intermittent) electricity. The concrete city was surrounded by more extensive slums, which it was possible for many of Luanda's richer residents to ignore. The Angolan elites and foreign aid and oil workers dodged potholes in air-conditioned Land Cruisers, and carried mobile phones that were then so chunky that they wore them with straps across their shoulders. The elites sat in restaurants on the Ilha and watched their sons, exempted from military service, race jet skis off the beach, while their daughters, perhaps exhausted from the previous night's cocaine-fueled dancing at the Pandemonium disco, sunbathed and sipped Brazilian caipirinha cocktails.

People sometimes talked of "100 families" who control Angola.[11] Unlike in Russia and eastern Europe, where democratic revolutions swept away the discredited old guards, the Angolan dinosaurs from Marxist-Leninist days adapted and stayed in control. In fact, many helped accelerate the headlong rush to petro-capitalism once they saw the fabulous opportunities that might come their way. For many of these people, Moscow and Lisbon were the centers of the outside world, but now, as oil production rose, their focus was slowly diversifying northward and westward: to Paris and London, then to New York, Houston, and Washington (and now, years later, eastward to Beijing).

José Cerqueira, a former senior central bank economist, said that corruption was not a big problem before 1990. "You did not really see these poor, miserable people," he said. "We went to Ethiopia and we were horrified at the hunger. Street children started appearing in Luanda in 1990; now what I saw in Ethiopia is here, in front of my house."

It was an age of lunatic accounting. Feisty women with babies strapped to their backs would sit in groups on street corners, hissing and beckoning at me, rubbing thumb and forefinger together. *Fala bem, Amigo!* Speak well, friend! These were the *kinguilas*, street moneychangers who greased the oil-charged economy and offered a "parallel" exchange rate, many multiples of the official one. For a $100 bill these loud-mouthed ladies would hitch up their skirts and produce a half-brick of kwanzas, wrapped in rubber bands. They were mysteriously efficient and beat the banks hands down: no queues or forms to fill in, they never cheated me or anyone I knew, and their rates changed from one moment to the next, in sync, all across town. The economic police periodically chased them away, but they were so important that they were never gone for long.

Once you had kwanzas, it was best to spend them fast. This was because of the way the oil system was set up. Sonangol's black box had properly emerged after the crash in oil prices in 1986, when a barrel of benchmark Saudi crude oil fell from $28 per barrel to below $13, and President dos Santos responded to the traumatic downturn by wresting part of the oil money away at source, inside Sonangol, putting it under direct control of the secretive presidential clique known as *Futungo* (after the name of the presidential palace),[12] and bypassing the ministries. "Until 1986, we . . . had reasonable control over the oil money," a former finance minister told me. "After that, some of it went directly to *Futungo*, and bypassed us. That is when the trouble really started."

As money disappeared down this presidential sinkhole, Angola's normal lines of credit, reeling from war and from $10 billion of foreign debt,[13] were all but dead. To pay salaries and to cover other obligations the central bank had to print kwanzas. Angola periodically paid De La Rue of Basingstoke, England, to send out planeloads of fresh kwanza notes, often with extra zeroes on their faces. When I arrived in late 1993, inflation was boiling at an annual rate of nearly 2,000 percent. As the value of the kwanza collapsed, salaries stayed still in kwanza terms, so people got poorer, and inflation wrecked the price signals you need to generate areas of efficiency in an economy. Oil did not exactly *cause* the inflation; if any one thing did, it was the war. But oil's strange set-up inside Sonangol helped split economic management in two, sending the accounts haywire and feeding the spiral of rising prices.

Were Angola's finances run with two sets of accounts: a fictional set to show to nosy outsiders, and a carefully guarded black book, to outline how the money *really* flows, down to the last penny? Or was it just disorganized chaos? I have never found out the answer, but I suspect it was half-way between the two.

Angolans with connections dreamed up ways to profit from the insane economic distortions. A British reporter who visited Luanda before I arrived gave one account of how he hatched up some free money.[14] The reporter needed gas for his car, but he was short of cash. So he paid 564 kwanzas, or $19, for six dozen eggs at an official *loja franca* shop. Next, he took the eggs to a market and sold them for 100 kwanzas *per egg,* yielding 7,200 kwanzas. He had started with 564 kwanzas, filled his car with enough gas to last him a week, and ended with over 4,000 kwanzas in change. Other people told stories of friends buying round-the-world air tickets for the price of a case of beer.

How could this be? You found part of the answer in a table of exchange rates in the daily newspaper *Jornal de Angola.* The key was to have the connections to be allowed to use the official rate, which meant that you could buy goods at subsidized prices. By early 1994 the *kinguilas* were offering 120,000 kwanzas per dollar, 20 times the official rate of 6,000.[15] A powerful official could then change $1,000 on the street, and get 120 million kwanzas in a shoulder bag from the *kinguilas.* With the right connections, he or she could then go to the central bank and ask for the official rate, at 6,000 to the dollar, yielding $20,000 of what was, in effect, free oil money from the bank. This was not normally taken in cash; instead, the central bank would pay it to a foreign supplier, who would deliver $20,000 worth of goods. The happy official would collect these from the port and sell the goods in the market, with another mark-up, and repeat the cycle.

Many other routes existed to tap into the cash flowing from the off-shore wells. A top official might borrow kwanzas from a state bank at low interest rates, convert them into dollars, then wait until inflation had eroded their kwanza value to peanuts before paying the loan back. So the banks built up debts, which the oil-fed government covered. The elites also obtained state assets in murky Russian-style privatizations, capturing the construction trade and banking services, which depended heavily on government contracts—paid, of course, with oil money. These myriad routes, which were all *perfectly legal*, drained the national budget. Meanwhile, children no longer got their schoolbooks or tetanus shots, and the potholes in Luanda's streets yawned wider.

It took time to grasp the depth of the subsidy mentality. In my hotel, a visiting scrap metal dealer was waiting for his dream deal: a contract to buy all the destroyed tanks that littered southern Angola. Mr. Pig Iron, as we called him, grew more and more exasperated over time. No matter how much he argued about world prices for iron and steel, and shipping costs, the Angolans still wanted him to pay fairy-tale prices. Any good business-man, they appeared to believe, had routes to his own government subsidies, so he would eventually accept a price that made no sense from a purely commercial standpoint. Mr. Pig Iron left Luanda in high dudgeon, leaving his hotel bill unpaid.

The subsidies and official exchange rates, which looked to outsiders like giant schemes for legalized theft, were officially justified as ways to get cheap food and goods to the masses. In practice, of course, the subsidies were captured not by the poor in the form of cheap food, but by big im-porters who wielded petty regulations against potential competitors, and so kept prices high. War protected their monopolies by keeping the roads closed, so competing local crops could not get into Luanda's markets. Apart from things like beer, soft drinks, or cement, which were too bulky and cheap to import profitably, and a few subsidized factory goods, nobody in Luanda really *made* anything. V. S. Naipul's novel *A Bend in the River* cap-tures some of the spirit of oil-rich Luanda:

> They didn't see, these young men, that there was anything to build in their country. As far as they were concerned, it was all there already. They had only to take. They believed that, by being what they were, they had earned the right to take; and the higher the officer, the greater the crookedness—if that word had any meaning. . . . I was nervous of getting involved, be-cause a government that breaks its own laws can easily break you.

The Peruvian economist Hernando de Soto, in his best-selling book *The Mystery of Capital,* argues that property rights are the bedrock upon which the edifice of capitalism is built. In Angola after independence, property was nationalized, chaotically. This ministry was given responsibility for that building, but nobody told them that another ministry had allocated it to someone else. Property rights became an ephemeral concept, a set of dull and shadowy ideas without real substance. When private property was reaccepted after 1990, the people who took advantage of the uncertainty were the established elites. A friend walked into his flat one day in Luanda—a nice apartment with a sea view—to find two soldiers sitting in it. It was theirs now, they said. My friend had a contact more powerful than theirs, and he got his apartment back.

Once the elites had extracted the main profits from the goods they had imported, they relaxed their grip, like dogs tired of an overchewed bone. Free market competition vigorously reasserted itself as thousands of poor street traders struggled, in an arena of fearsome price competition, for the last few crumbs of profit from these imported goods. The heart of this desperate capitalist world was Roque Santeiro, a vast street market that was named after a wildly popular Brazilian television soap opera, and which perched on a long garbage-strewn cliff above the port. (Today the market turns over well over $5 million a day.) You could buy anything in Roque: amid imported Brazilian chicken meat and Portuguese pigs' trotters I spied expensive Scotch whisky, a London A-Z street map, and a soldier hawking ammunition. I saw cinemas showing Kung Fu action films, rows of coffins for babies, raucous bars, a terrifying *dentista popular* with Mad Max dental equipment, and sinister dwarves performing unwholesome tricks for money among the sewage. I saw men going in to the Casa de Pornografía for $2 congress with whores on a stained blanket, in what had to be the world's filthiest brothel. I declined a policeman's urgings for me to have a go. According to one rumor, which was probably untrue but credible enough for people to believe it, a thief stole a container from the port and opened it in Roque, to find a MiG–23 fighter-bomber, without its wings.

Many importers who fed Roque were Lebanese, serving as contractors to rich Angolans who had access to oil money to buy imports. The Lebanese succeeded because they were good at sourcing foreign goods cheaply, and because they had extensive networks of trust, which made up for the generalized lack of it in Angola's business world. I coaxed one of them to talk:

I came in 1992 during the fighting, and there were bodies in the streets. Nobody else wanted to come, but the market was so thirsty you could sell goods at triple the price. It was like the Wild West: you brought 100 containers and 20 were stolen. The other 80 took a month to get through the port, but those 80 would make you rich. People who stole containers got even richer—they did not have to buy the goods in the first place. In Roque there are no contracts, everything is done on trust—so we don't sell to just anyone. We sell to two or three big *senhoras,* and they sell them on down. If you bring a container with good prices, it's like war outside, you need armed guards. You get kwanzas from the market, you change them through the *kinguilas,* and you take the dollars to the airport—or you can change the dollars for diamonds, which are easy to smuggle. You give the guy here diamonds, and he arranges for the dollars to be transferred to your account overseas; he smuggles them out himself. Roque is a good washing machine: you can launder your money here. So many deals have happened between us [Shi'ite Muslim Lebanese importers] and the Israeli diamond dealers. Forget about religion or politics: here it is just deals. Angola's prices are set here. People selling in Huambo buy here. The price in Huambo is the price in Roque, plus transport, plus the trader's profit. Roque is like the stock exchange for Angola.

The market had an intricate hierarchy and structure, which had sprung up spontaneously and was directed by nobody. Portuguese biscuits might be imported by one group, brought through the port bureaucracy by another, loaded by a third into a truck, which is driven by a fourth, and unloaded by a fifth. The truck is owned by a sixth; another group will buy the biscuits wholesale then sell to retailers. *Kinguila* money changers and even diamond dealers might operate between the importer, the wholesaler, and the retailer. Though six people did six separate transactions, one could probably do them all. These unwritten social regulations say who can do what, and this creates the jobs and helps prevent poverty from turning into riots.

As we went about life in the Luanda bubble, the fighting rumbled on elsewhere. We journalists pestered aid workers for snatches of news. It was strangely easy, sometimes, to get into the interior. I would drive to the airport and wander on to the runway to ask the pilots where they were going. Once, a beery Ukrainian crew agreed on the spot to fly me to a desolate government-held town in southern Angola, where some aid workers were feeding some refugees; then they flew me back the same day.

The foreign press corps consisted of memorable characters. One was a bearded Croat who had remarkably good army connections. There was Vladimir: a wiry, coiffed young Russian who worked for the Tass news agency, an astute but jaded observer. "To my mind, this story is shit," Vladimir would announce, having noted, and seen through, the latest improbable government communiqué. His cynicism was a delight at times, but often his mood descended into a beery melancholy. Marco, a Belgian photographer and a fabulous raconteur, took some of the best war pictures I have seen, yet he lacked ambition. My pick of the bunch was Chris Simpson, the BBC's resident correspondent who lived next door to me in my hotel. Chris was a walking encyclopedia of British sport, music, and television trivia, and was a charming, if somewhat clumsy, man. He had been known to trip in the dark into open manholes into sewage ("more of a swamp," he later insisted); once he was tried in court (and acquitted) because he had, through a train of unlikely mishaps that began with some beer at a football match, ended up caught by police *inside* a prison cell holding some big-time South African cocaine smugglers. As he waited in prison for a week before his court case I naively sent Chris pizzas and a copy of Nick Hornby's football novel *Fever Pitch*. His cell mates stole the pizza and used Hornby's excellent novel as toilet paper.

Foreign correspondents would fly in from Johannesburg, thrilling with adrenaline. A photographer called João strode up and down the length of the press center bar, swearing. "I need to get to the fucking front lines and these fuckers are trying to stop me!" he yelled. Abraão's reports had infected everyone. We all wanted to get to Kuito.

By the middle of 1993, Kuito had become Savimbi's punching bag. If the army won a victory against UNITA somewhere, he took it out on Kuito. Dogs were eating bodies in the street, but by now its starving residents were eating the dogs, too, along with cats and rats. Injured patients in soiled bandages crawled from hospital beds to chew on grass and flowers, and old women lost their legs as they searched minefields for cassava. Corpses filled the river and soiled the drinking water, bringing dysentery. Adults and children deflated slowly down onto their skeletons. By now, perhaps 35,000 people had died in nine months of siege,[16] leaving about the same number alive. Abraão's voice, recounting the latest UNITA "flagellation," sounded hysterical at times.

Yet the first signs were starting to emerge that the pendulum of war, which had swung a long way in UNITA's favor, was changing direction. None of us journalists knew it then, but new weapons were being unsheathed.

In January 1993 UNITA had overrun the northern oil base of Soyo, where the Belgian company Petrofina, along with Texaco, Elf, and others, had big onshore operations supporting nearby offshore wells. UNITA had lobbed warning shells into the sea near some of the close-to-shore rigs, shutting down about 7 percent of Angola's oil production capacity (the rest was out of reach, farther offshore.) A small oil company put the Angolans in touch with Executive Outcomes, a new mercenary outfit based in South Africa that was cofounded by Simon Mann, an upper-class Briton (who would, more than a decade later, be embroiled in a famous failed coup plot in oil-rich Equatorial Guinea nearby). Executive Outcomes was a modern incarnation of the mercenary adventurers who have swaggered across Africa in years past: the fabled Frenchman Bob Denard who operated out of oil-rich Gabon and specialized in overturning governments, the Irish-born Mad Mike Hoare who once fought in the Congo, and some incompetent British and American mercenaries who once fought alongside another (now-defunct) U.S.-funded Angolan rebel movement. Executive Outcomes recruited men from the former South African apartheid army, many of whom had once fought on UNITA's side—and with salaries of between $2,000 and $8,000 a month, there was no shortage of volunteers. They knew the terrain, and the enemy, intimately.[17] They recaptured Soyo in February 1993,[18] giving the oil companies time to haul out expensive equipment before UNITA recaptured it. Three months later, Executive Outcomes took a phone call from an Angolan general, who asked them to train the MPLA's ramshackle army.

Yet President dos Santos was about to unveil an unusual new oil weapon. The oil was flowing too slowly to buy enough arms to turn the war around quickly, and western banks were unwilling to lend money for arms to a recently Marxist-Leninist government that was then under an international arms prohibition, and whose enemy still had allies in Washington. So Angola did two things. In April 1993, it unilaterally repudiated the arms prohibition that it had signed up to as part of the now-collapsed peace agreement. Also, President dos Santos contacted the formidable French Africa networks in Paris.

By then, Kuito was plumbing new depths of savagery. In September, after intense international pressure, UNITA let the United Nations fly in some food, under an unholy bargain: it was enough for 100,000 people, though Kuito's defenders numbered 30,000. A bit was stored but the rest went to UNITA—enough to feed most of its army. "It was not a perfect arrangement," a U.N. official said. "But it was the best we could do." A U.N. official who went in with the food said the town was like a scene from Dante's *Inferno*. The fighting started up again, the food flights stopped, and the hunger returned.

In November, President dos Santos's French contacts cooked up an oil-backed loan, worth $47 million, which was then used to pay for an arms shipment.[19]

None of us knew it, but an inflection point had been passed in the Angolan war. Angola and France had also opened a door to an episode with huge international repercussions, whose effects are still being felt today. Two weeks after the arms deal, UNITA and the government began peace talks. The mood lightened a little, and in January 1994 UNITA gave me permission to visit Kuito.

The U.N. aircraft came down in a tight spiral to avoid antiaircraft fire, and just after we landed a heavy machine gun opened up past some eucalyptus trees, then died down as a temporary ceasefire took effect for our visit.

Abel Abraão's home sits on low hills at 6,000 feet altitude in Angola's central highlands. With refreshingly clean, cool air, this had once been a jewel of the Portuguese empire, a center of agriculture and industry at the heart of Angola's most fertile zone. Kuito grew up as a staging post on the giant East-West Benguela railway, which British engineers had built for the colonists.

Now Kuito was ruined. Houses lay uprooted, with delicate arches and elegant Portuguese mosaics strewn in pieces on the street. Four story structures lay collapsed as if crushed by giant cannonballs; apartment blocks had their faces shorn off; bullet holes and the splash patterns of rocket-propelled grenades peppered once-beautiful pink walls. Each street corner presented an exhibition of the grotesque. One, near the hospital, resembled a colony of giant frogs, with squads of one-legged mine victims, some armed, hopping about with homemade crutches. On another corner, small boys proffered damaged arms, legs, and heads. "Feel my shrapnel wounds," said tiny, impish Bruno, scampering along beside me, patting his head and guiding my hand to a ridge on his scalp, which, he proudly assured me, still had metal inside. In the hospital I could see a girl's teeth through a two-inch hole in her cheek from an untreated ulcer. Nearby, pupils in a classroom studied chemistry with their AK–47s stacked outside; in the road a red-painted circle warned of an unexploded shell. An attentive man walked behind me for much of my visit to the world's most heavily mined town. *"Anda bem,"* he would say, guiding me firmly away from inviting-looking shortcuts. "Walk carefully."

UNITA had sliced Kuito into four: a square mile or so of the town proper; the airport; the army barracks; and a position on a nearby hill. The weeks turned into months as UNITA attacked, and as the hungry defenders of stubborn little Kuito clung on in these crazy pockets. Some of the town's residents lost their minds. An Angolan television crew filmed one injured lunatic pulling rotting flesh off his own leg and eating it. Residents traded with their ethnic UNITA brothers: nuts, cassava, or onions from UNITA, in exchange for clothes, salt, or cigarettes that came in the parachutes. This was not their fight, but their leaders'.

Four timid teenagers, shouldering AK–47s, described the terrifying *batidas,* when men and women—sometimes 150 at once—sneaked out at night into UNITA-held land on expeditions to forage for potatoes, maize, goats, and firewood. Before returning with food, often days later, they would listen for Abel Abraão on the radio, to check that Kuito had not fallen in their absence. Sometimes they had to fight their way back in; on some expeditions, only half returned alive. During the siege, government Ilyushin cargo aircraft at 10,000 feet parachuted weapons, food, and salt down into Kuito's besieged pockets. The packages often fell in UNITA territory, so groups of defenders would have to fight their ways out to them and drag them back, under fire. I never got to see Abel Abraão on that trip: UNITA had a price on his head, and he stayed out of sight.

After we left, the fighting started up again, and UNITA rained its diabolical mortars down onto Kuito again. The defenders doubted that they would survive for much longer. As Abraão kept assuring us in his reports, they had all vowed to fight *até as últimas consequências*—until the end.

In April 1994, the French networks cooked up a second, much bigger, secret oil-for-arms deal, worth about $500 million. Russian-made tanks, rockets, helicopters, troop transporters, and combat vehicles began to arrive.[20] You could see that the army had fresh weapons. In one town near Luanda I met a group of young soldiers in a rainstorm. They had been trained by Executive Outcomes, and they were sporting the latest AK–74[21] rifles, with evil-looking attachments that I had not seen before. These youths, very apprehensive, were about to head out and capture the strategic road junction at Ndalatando, and they would then join the mercenaries in a victorious assault on rebel bases in the northeast, where

UNITA mined the diamonds that funded its war effort. With the oil weapons unleashed, the government began to roll the rebels back.

For a while, Kuito remained in bloody suspended animation. The combatants would fight, then a temporary calm would descend and the United Nations would fly in some food; then the fighting started again. One particularly vicious episode broke out when some soldiers tried to drag away a fallen tree branch, and UNITA tried to stop them, first firing warning shots in the air, then firing right at them. Before long, mortar and artillery were once again raining down.

Along with Karl Maier of Britain's *Independent* newspaper and Alex Vines of Human Rights Watch, I visited once more. The town was even more wrecked than the last time. It was ceasefire time again, and soldiers from both sides were—tentatively—chatting across one of the front lines. There were even stories of bicycle races between the enemies. The cornered residents were doing everything they could to preserve a sense of normal life. There was a well-attended driving school, though nobody except aid workers had, or dared to use, cars. Policemen had even tried to give the aid workers parking tickets, though the idea of parking spaces had lost any meaning amidst the rubble. Teachers were working for no pay, and classrooms were full. We saw a long column of UNITA women snaking down a nearby hill carrying firewood on their heads, to trade with the defenders. "This is a war of madmen," said a businessman, whose shop now housed refugees. "One moment they trade, and in the next, they kill each other." He had recently saved one woman, who was lying injured in the street in view of the UNITA snipers, by throwing out a rope and dragging her in.

We wanted to stay overnight, but the United Nations suddenly said that we had to leave immediately. On the airport road a sinister, grinning UNITA captain told us that Kuito was about to fall. Straight after we left, hell broke loose.

For months, the defenders had been stockpiling arms from the airdrops, waiting for the final battle. When the command came, the defenders, emerging from their separate pockets of territory, crashed into the UNITA positions, scattering the rebels. "Everything was exploding," said Popi, a civil defence soldier who described the battles to me years later. "Snipers were shooting down the main street, bullets were hitting everywhere. They were shooting men, women, children, dogs. This was our last chance, and we gave everything we had. Our martyred town rose for one last battle. It was as if the dead had risen out of their graves to fight alongside us."

In a few days, the defenders smashed UNITA out of the suburbs and into the bush. At the end of June I switched on my radio to hear Abel Abraão, broadcasting from Kuito airport for the first time since the siege began. "The barricades that divided our city ceased to exist from yesterday," he said exultantly. "We have captured the diabolical mortar . . . the army controls the whole city." The tide in Angola's war had turned. By November UNITA had signed a peace agreement, on humiliating terms.[22]

After the war, I stayed in Luanda for several months. In those days everyone believed that Angola's poverty was the legacy of war and Stalinist policies. There was little academic research then about the harm that oil can do. Angola, diplomats told me, would now harness its oil and use it to build an affluent African exception. There was probably not a day when I did not think about Kuito.

The town briefly knew world fame in 1997 when Diana, Princess of Wales, walked in body armor through a Kuito field recently cleared of landmines. I wondered then what had happened to Abraão.

It was six years later when I finally tracked him down.[23] The American oil company Chevron had recently named a big new oilfield "Kuito," but—though the town's police station boasted a big satellite dish, and impressive vehicles were parked outside the ruling MPLA offices—the wealth seemed not to have reached the town. It still lay in ruins.

Abraão lived in an appealing middle-class colonial residence near the town center and he seemed surprised to get a visitor. He invited me in for tea and biscuits. Children clambered over him as we chatted about how the fighting had started. "I had just come back from journalism training in Cuba," he said. They knew UNITA had been reinforcing their positions. One day a general came to see him. "He told me that the war starts tonight. All I had was this," he said, patting a gray bakelite rotary telephone by his sofa. "They had been killing MPLA sympathizers and throwing them down wells. If UNITA did this, we thought, then it was better to die fighting.

"The resistance was pure bravery—that was all we had! We knew we would die here—especially me: I was the voice of Kuito. We were afraid, yes, but the fear went away. They put a price on my head: $300,000, dead or alive. At night, they shouted, 'Attention, Abel Abraão! In one or two hours, you will die! We will cut you in pieces.'" As he said this, he made chopping movements with his fingers down his outstretched forearm. "They sent beautiful women in to find, seduce, and then poison me."

"My house was on the frontline and UNITA were in that building, across my garden," he said, pointing out the back kitchen. "They never knew I was here, right here." One night UNITA raided his home on a foraging mission—he showed me the place where the lock on his back door was shot through, and where several bullets had drilled into the side of his refrigerator. He was out at the time, and they took everything.

There was not much more to say; he has done this interview often before. He has settled down to a rather pedestrian life as a middle-ranking correspondent for his old state radio, in what is now a quiet town, with fading memories. As I left, he showed me a dark blue Hyundai Elantra sports car, which President dos Santos had presented to him after the war. He hardly drove it, he said, because of the potholes. The token of his heroism lay in his garage covered in a tarpaulin, slowly gathering dust.

4

OMAR BONGO

TAKING THE RED PILL

After leaving Angola I was traveling in French central Africa when I got an unsolicited e-mail from Alain Autogue, a Frenchman who had heard from an editor at the *Financial Times* that I was to visit Gabon. He said he wanted to help me.

I knew little about Gabon except that it was a quiet, oil-rich former French African colony whose people consumed more champagne on average than anyone, including the French,[1] and that its diminutive president, Omar Bongo,[2] kept popping up in parts of the world that did not obviously concern him. A couple of years earlier, there had been newspaper reports involving Bongo, some call girls, and an Italian couturier, but Gabon had then, like a crocodile in a swamp, sunk quietly back out of sight. French newspapers were reporting Gabon's role in some judicial investigations in Paris, but the stories were too complex to make much headway in the English-speaking media.

Looking back, Mr. Autogue's invitation reminds me of a moment in the science fiction film *The Matrix*, when the main character is presented with a blue pill, a route back to normal life, or a red pill, offering an alternative reality, a chance to "see how deep the rabbit hole goes." I asked him: What did he have in mind? An emphatic message came back from Paris. "Someone will pick you up at the airport. Don't worry. *Bon Voyage.*"

It was obvious he wanted to influence what I would write. I was suspicious about French people in Africa after an incident a few years earlier when I had run out of money while backpacking in French-speaking Ivory Coast. I took my credit card into a bank, where the clerks sent me upstairs to see a big-boned Frenchman behind a large, paper-strewn desk. He told me my card was in arrears, and pulled out scissors and, with Gallic insouciance, cut it in half before my eyes.

Yet I was curious, too, so I accepted Mr. Autogue's invitation. I took the red pill.

I had left Angola two years earlier for London, where Reuters retained me as a reporter on the oil and gas desk. The job involved ringing up traders to try to follow price fluctuations in refined products from crude oil: heavy fuel oil for ships and power stations, lighter diesel and gasoline for vehicles, and naphtha and liquefied petroleum gas for petrochemicals and other uses. Price movements in these markets are complex: every oilfield has its own character, and refiners prefer West African crudes to Middle Eastern oils because they are mostly sweeter (fewer impurities like sulfur) and lighter (you can squeeze out more gasoline when you refine them). The prices of crude oil and its refined products all flicker up and down absolutely, and relative to each other, minute by minute, responding to weather, geopolitical tensions, and many other factors. Traders constantly try to manipulate the markets in their favor, and to stay ahead of their competitors they gather detailed political and economic intelligence about the countries that sell them oil. Western governments sometimes help them; sometimes it is hard to know where the world of spies ends and the world of oil begins. The job taught me that the oil trade is, as much as anything, about information: whoever knows the most makes the most money. The huge complexity of energy markets, and the benefits that market participants can derive from their secrecy, partly explain why it is so hard to bring about transparency in the oil industry. Corruption abounds.

I disliked the job, and my raw, fresh memories of Angola made life in London seem gray and superficial. So I left in 1997, planning to work as a freelance journalist in beautiful Cape Town, from where I could return to Angola. Gabon was a stepping-stone between London and Cape Town. Just before I left, a commercial intelligence company had contacted me, offering a lot of money for a broad analysis of Gabon and its prospects, plus a who's who of the country's most powerful people. They would not tell me (and I still don't know) who the client was. It was a lucrative, if curious, start to my freelance career. I was subsequently contacted by a similar firm, with a rather nastier proposal: the upper-class British voice at the end of the phone wanted me to go to Cameroon and *pretend* to be a journalist, to assess the "strengths and weaknesses" of a certain government minister. Another anonymous client—probably a creditor looking to seize Angolan financial assets overseas—promised me "really very large sums indeed" if I could trace bank accounts where the Angolan central bank stashed its money. These projects seemed not only wrong, but dangerous, and I rejected them.

Soon after Mr. Autogue's e-mail, I flew to Libreville. At the airport a white man in aviator sunglasses met me and helped me carry my bags to a red sports car. He was a French statistician working with the Gabonese government, and a business partner of the mysterious Mr. Autogue. We drove south down the Boulevard de l'Indépendance, a six-lane coastal artery clotted with red-and-white taxis, Mitsubishi Pajeros in polished chrome, and police wagons with scowling men with black jumpsuits and pump-action shotguns sitting on benches in the back. To the right ran a beach with palm trees, old canoes, and black and white joggers; inland from the road we passed high whitewashed walls hiding wealthy homes, which were framed by clean tarred roads leading up a shallow incline away from the sea. Near the town center the buildings grew taller and uglier in a progression that culminated in the vile Palais de bord de Mer, a menacing hulk of gray cement that is President Omar Bongo's main office. We passed the town center and the commanding Elf tower by the estuary, then, after the road had narrowed and became more ramshackle, we got to my hotel. I was told that Mr. Autogue would join me the following morning.

I hired a taxi to go exploring. Gabon straddles the Equator, between Cameroon and Equatorial Guinea to the north, and Congo-Brazzaville to the south. It has fewer than one and a half million people, a third of whom live in Libreville. Gabon is slightly larger than Britain, with a fortieth of its population, so there is space for the inland oil rigs, and reports of friction with local communities are rare. Spills do happen,[3] but with few witnesses and an obsequious media, it is easy to hush them up. Inland from central Libreville the houses get smaller and soon you reach neighborhoods filled with sewage, roaming piglets, corrugated iron, and children scampering among papaya trees. The city then peters out and, after some patchy agriculture, you enter a forest as big as western Europe, which extends far across the borders and contains nearly two hundred mammals, including silverback gorillas, forest elephants, and hippos that swim in the Atlantic surf. This forest hosts the deadly Ebola hemorrhagic fever, as well as iboga, king of hallucinogenic plants, which offers a route to the land of the ancestors and has a profound cultural importance here. On geologists' maps the oilfields, which started pumping in the late 1950s, are splashed like flecks of paint across this forest and out to sea.

Libreville is a bit like the capitals of some Gulf Arab states. The working people who lug bricks or bend metal are usually foreign Africans: Ghanaians, Congolese, or Togolese, who collectively make up a quarter of the population. Like Pakistanis or Palestinians in Arab oil states, these foreigners are sneered at and mistreated. I saw a foreign taxi driver, stripped to his shorts, being forced by Libreville cops to do push-ups in the road. When oil prices dipped in the mid-1990s, Gabon expelled 55,000 foreign Africans, though they soon trickled back for the oil money. High wages also attract child traffickers who bring poor kids in from Togo and Benin and sell or rent them as house servants or prostitutes. Oil, which accounts for 80 percent of exports (the rest then mostly being manganese and timber), and a cloying French civil service tradition have fostered a sleepy bureaucratic lethargy among many Gabonese, who consider menial activities beneath their dignity. They mostly aspire instead to be among the 50,000 or so Gabonese idling in the oil-fattened civil service,[4] or even to break into the exceedingly wealthy top elites. "The problem," the foreign minister later told me, "is that many of us Gabonese just do not want to work."[5] Atop the expatriate pile perch nearly 10,000 French, who shop in supermarkets for croissants and imported gourmet cat food; they and the Gabonese elites shampoo their poodles, employ foreign Africans to clean their swimming pools, and, in late November, wash down Parisian cheese with the year's first Beaujolais wine.

President Omar Bongo towers over his country, unchallenged. When he came to power in 1967 at the age of 32, he was the world's youngest president, and today he is Africa's longest-serving ruler. With twinkling eyes and a neat moustache, he is short (and apparently self-conscious about it, sometimes wearing platform heels to compensate), impetuous, and often irascible, reportedly flying into rages when his, or Gabon's, interests are threatened. He lets local private media be critical, within limits. Human rights organizations find relatively little to complain about in Gabon, which is less brutal than most countries in this troubled region. Bongo claims to have no political prisoners, and he tolerates a skewed kind of multiparty politics. He no longer wins elections with over 99.5 percent as he did in the 1970s and 1980s (later calling these results "perfect reflections of the truth"). In two elections since 1990, his share of the vote collapsed, to less than 80 percent.[6]

Bongo's big splash recently was in a front-page *New York Times* story about Washington lobbyist Jack Abramoff, who reportedly asked Bongo for $9 million to arrange a meeting with president George W. Bush. Bongo has denied ever meeting Abramoff or paying him anything.[7]

He is also reported to have a healthy sexual appetite. "If you see old French guys at the airport accompanying attractive women, they are probably headed for him. His ex-wife is in a mansion in Beverly Hills now—there was a time when she would chase him around the palace with her shoe in her hand," said a foreigner who knows him. "He is a pesky little feller." There was a diplomatic row in 2004 when a 22-year-old Peruvian beauty queen was invited to a pageant in Gabon. Soon after arrival she was ushered into a paneled room, where Bongo was waiting. He pressed a button and some sliding doors opened, revealing a large bed. She fled. A trickle of articles like this means that the rare news coverage about him in the English-speaking world portrays a clownish, sexually voracious little despot, otherwise of little interest. A story about the beauty queen, for example, was entitled "Beauty and the Bongo."[8] Another article called him "an obscure thug . . . an extraordinarily rich man."[9] An American magazine described him as "the dictator of Gabon . . . a tiny, natty man, very black," accompanied by a queenly wife in Gucci sunglasses and lashings of gold jewelry who towered over her petite husband, who was "exceedingly corrupt and exceedingly wealthy."[10]

These kinds of stories totally misrepresent Bongo. He's much more than this. He's not obscure, and of all the African leaders in the region, he's possibly the least thuggish. And, after all, what is corruption? Many Africans would see what he does, judiciously allocating resources to shore up his power, as entirely normal in the African context, even as an essential tool for survival. Of all Africa's leaders, he fascinates me the most. He is a political magician who has projected secret influence around the globe with a force far, far out of proportion to his country's tiny population. In France, for decades, he has long known everyone who is anyone, and a few more besides. France's top politicians take his phone calls personally; Bongo has had more influence over French politicians than they would admit, and he is smarter than most of them—which, given the strange, quasi-meritocratic French political system, is saying something. On his regular trips to Paris he prefers the Hotel Crillon, spending thousands of dollars per night, and when word gets about that he is in town, the foyer fills with furtive lobbyists, businessmen, and politicians who come to wheedle favors from the little big man. Bongo remembers everything and everyone he meets, and he has helped broker secret agreements between big political factions in France. If you look carefully, you will see Bongo's fingerprints all over past peace deals signed between governments and rebel movements around Africa. Pictures in his offices show him meeting all the world's top leaders, from Mao, to Nixon, to Mandela.

He was born Albert-Bernard Bongo in a rural village near the eastern border with Congo, the favorite son of a farming village chief and the youngest of 12 children. His father was polygamous; according to Bongo, celibacy was not condoned because the community tended to treat with suspicion men who were less than susceptible to feminine charms. They lived in adobe huts with straw roofs, and as a small child he ran around naked or in a raffia loincloth. "Nobody questioned the head of the family. The men did the heavy work like plowing and went hunting; the women planted manioc and fished. When you came back from hunting or fishing, you shared the catch among everyone, starting with the handicapped. No one was left out."[11]

When he was seven his father died, and one of his brothers took him away to what is today's Congo-Brazzaville in French Equatorial Africa, a French colonial territory that also included today's Gabon, Cameroon, Chad, and Central African Republic, and where Charles de Gaulle organized French resistance and set up his capital of "Free France" in World War II before becoming French president in 1945. Bongo's father, just before he died, entrusted the family's ancestral artifacts and secrets to Bongo, since his eldest brother, who should have received them according to local custom, was away. These gave Bongo a special role in the family giving advice, and settling disputes.[12] This role surely prepared him for leadership in later life.

When World War II ended, Bongo went to school. He was bright, and was selected to be sent to France for further education, but his older brother refused to let him go. Fuming at the missed opportunity, Bongo became rebellious; once, according to a memoir, he grabbed a cane from his teacher's hands, and whipped him. He was chased out of school, and left town. At his next school, and at college, he organized strikes and protests, and on a visit to France he was inducted into the Freemasons.[13] At some point, the French colonial authorities began to take an interest in this charismatic, intelligent young socialist agitator. Nationalist sentiment was simmering, and as independence approached, it was essential to identify the future leaders.

For Britain, independence had been a matter of handing the colonies over to Africans, with as little fuss as possible. Britain was not without its paternalistic streak: premature independence, a British minister had once warned, would be "like giving a child a latch-key, a bank account, and a shot-gun."[14] Nevertheless, many people in Britain, which had already granted independence to former colonies in Asia, saw the idea of eventual independence for the African colonies as a natural inevitability, and sought to build up universities, infrastructure, health, and education before handing over.

France had a more intimate (you might say incestuous) relationship with her colonies. They were not considered separate territories so much as they were considered *part of* greater France, and the idea was to bind them tightly to the mother nation. "France wanted to buy time during decolonization," Bongo said. "It offered French nationality to many young Africans. The idea of becoming a French citizen enchanted us. We were going to live like the whites!"[15] Some Africans could even elect representatives to the French parliament directly: a Senegalese became the first African elected to the French parliament in 1914, and he rose to become a French junior minister. Others followed.

The African colonies also underpinned what many in France called (and still call) the "French cultural exception," and what President de Gaulle called "a certain idea of France": notions that embody a French worldview steeped in sepia images of wise philosophers, rustic winemakers in the Loire valley, and toiling peasants in Provence. These visions have been a comforting refuge for French people over centuries, in the face of British imperialism, a rising Germany, then the Soviet menace, and now expanding American might. Exceptionalism inspires French foreign policy and provides its leaders with a clear imperative: France must punch above her weight in the world. Carefully nurtured African hatchlings like Gabon, helping France make her case for permanent membership of the United Nations Security Council after World War II, would continue to help France project her language, culture, and prestige abroad. The colonies were a secure source of strategic minerals, of profits for her firms, and of markets for French manufactured goods. Independence threatened to upset this applecart.

If the main threat to French interests was from the Soviet Union, there also was a secondary threat (which in many French minds was the primary one): predatory American companies, to be kept out at all costs. By the late 1950s, as independence neared for many French and British colonies, oil was starting to be discovered in Nigeria, Gabon, and Angola, and the colonial powers were staking out huge oil exploration licenses for their companies—the modern equivalent, if you like, of the flag planting by European explorers in the early days of empire as a way of asserting claims to new colonies. These licenses, embedded in international law, would be potent packages enabling the Europeans to keep a large degree of control after decolonization.

For all the naked self-interest, French Africa policy had pragmatic, noble strands, too. "At independence France supported and accompanied these African countries toward sovereignty," the French superspy Maurice Robert wrote in his memoirs. "Without this, their births would probably have been bloody, with terrible interethnic conflicts. This was much better than just letting them find their own way after independence."[16] There is surely something in his words. Indeed, President Félix Houphouët-Boigny of Côte d'Ivoire, in a speech before independence for his country, couched independence as "the challenge of the century," a grand experiment to see whether Britain's laissez-faire decolonization, or France's more intimate, controlling version, would turn out superior.[17] As a committed Francophile, he fondly hoped the French model would triumph.

As independence approached, the French secret services closely watched local agitators like Bongo, who had already made eight clandestine trips to Mao Ze-dong's red China. "Dangerous element," Bongo later discovered in his security file. "Watch him closely."[18] The story he tells about how he obtained his file is revealing, too. In 1958, French Equatorial Africa was being split into different countries, including Gabon, ahead of independence. One day Bongo, who had got a temporary job in a telegraph office in Brazzaville, noticed a telegram from a French general with instructions about which leader was going to be made to win which election. Bongo was shocked, and leaked the information to local newspapers in Brazzaville, which feasted on the story. He was arrested and charged with revealing professional secrets, but he was acquitted because, as a temporary worker, he had not signed any secrecy clauses.[19] After that, Bongo used his contacts in freemasonry to get a job in army intelligence. Once in, he donned his uniform and went to police headquarters and asked for the files on all the subversives. He found his own, slipped it under his clothes, and walked out. "They tried to find my file," he said with amusement, "and they never did!"[20]

Independence came and went in 1960, and France installed as the president Léon Mba, from the largest tribe in Gabon, the Fang, which spreads across the borders into Equatorial Guinea and Cameroon. Three years earlier, Gabon's first oil cargo had sailed for Le Havre in France, and exploration prospects were promising. With a fifth of the world's known uranium (Gabonese uranium supplied France's nuclear bombs, which French president Charles de Gaulle tested in the Algerian deserts in 1960),[21] big iron and manganese deposits, and plenty of timber, Gabon was already attracting unusual attention at the Élysee Palace of the presidency in Paris.

President de Gaulle dreamed of breaking the hegemony of the hated Anglo-Saxon "Seven Sisters," the forerunners of today's ExxonMobil, BP, Shell, and Chevron, and wanted a French champion or two. While one French company (what would become Total) focused more on North Africa and the Middle East, de Gaulle set up another company, later to be known as Elf Aquitaine, with Gabon as its springboard. Special contract terms for its subsidiary Elf Gabon, negotiated with compliant Gabonese rulers, would help the parent company catch up fast with its international rivals. Later, Bongo himself put the relationship neatly. "Africa without France is a car without a driver," he said. "France without Africa is a car without petrol."[22] Elf would become French President de Gaulle's strong arm in Africa: a vast offshore slush fund for channeling money secretly around the world, helping bend foreign leaders to his will, and an effective weapon against American and British companies competing with the French industrial giants. "De Gaulle wanted a company under full state control, his secular arm in the oil world, to affirm his African policies," a subsequent head of Elf later explained.[23] "Elf is not just an oil company but a parallel diplomacy to control certain African states, above all at the key moment of decolonization. Alongside exploration and production, opaque operations were organized, to keep certain countries stable."

President de Gaulle's schemer-in-chief was Jacques Foccart, a master manipulator who emerged from the shadow world of the French Resistance during the World War II.[24] Having carried out undercover operations in Nazi territory—smuggling arms and money in, and Jews and spies out—the former Resistance fighters had developed considerable skills in building clandestine networks that they deployed to project power and influence into enemy territory. After the war, de Gaulle had asked Foccart to deploy his formidable networking skills in Africa, to help keep the colonies bound to France. Known as the White Sorcerer, Foccart spun intricate cat's cradles of networks, or *réseaux*, to cocoon the former African colonies, and his spies rotated in and out of key posts in the secret services, diplomacy, and Elf Gabon. The relationship was underpinned by secret defense accords, masterminded by Foccart; in his desk he kept letters requesting French military help, presigned by friendly African presidents; he needed only to add the date. These accords—a bit like deals America has made with Saudi rulers—protected the African leaders, in exchange for free rein for French companies in their lands.[25]

Today you can see the remnants of this relationship off the airport road in Libreville. There is Bongo's mansion, preceded by Roman portals, tall black gates, and a long driveway. Next door, linked to it by underground escape tunnels, there are cement buildings and half-pipe hangars with radio nests and white men with military fatigues and short haircuts. This is Camp de Gaulle, a military base with several hundred soldiers from the French sixth marine infantry battalion, successors to the old colonial troops (the marine infantry is still sometimes called "la coloniale.") They drive Renault Clios into town, shop in the supermarkets in full uniform, and drink and pick up local women in the bars. They also deter would-be plotters from scheming to topple Bongo.

With a fox's instinct for the shifting currents of power in Africa, Foccart ran networks so effective that America and Britain grudgingly accepted France's mighty grip, which robustly kept the Soviets at bay. In 1965, the CIA asked the French secret services what was going on in Africa, to help coordinate the fight against communism. "The Americans were floundering in unfamiliar lands and cultures," said Maurice Robert, Foccart's top spy in Africa (he later worked for Elf too), who was asked to lead the collaboration with the CIA. He said the Americans (in a preview of American intelligence mistakes in Iraq four decades later), "accepted as fact the fantasies that opposition leaders spun them. . . . They asked me a thousand questions . . . the Americans had an abyssal ignorance of African affairs."[26]

For France, it was crucial to have the right people in place. Gabon's new president Léon Mba was so pro-French that he had proposed, ahead of independence, that Gabon should not be independent but should stay part of France; for a while after he took power, independent Gabon's flag had a little French flag in one corner. The Gabonese ministries and state companies had discreet white men toiling away inside them, pulling the strings. Foccart also had a plan B: the very young and very ambitious Albert-Bernard

Bongo, who was shinnying fast up Gabon's political ladder. "This lad was serious and intelligent and was destined for a great future," wrote the spy Maurice Robert. "He was a tireless worker who impressed me with his cool head, his aptitude to analyze situations quickly, and to act."[27] Bongo was, like Mba, excruciatingly pro-French, and deeply suspicious of America and Britain. He has remained so ever since. "All those in Africa who speak French should mobilize to defend their values, to affirm their presence, if not to say their superiority," Bongo said recently. "The English speakers stick together. They have formed a block against us."[28]

President Mba struggled to make the transition from African chief to statesman, and also found it hard to manage Gabon's seething ethnic divisions. To Foccart's consternation Mba became increasingly despotic, publicly flogging people who irritated him. In 1964, not long after the discovery of the great Anguille Marine oilfield, some Fang soldiers launched a coup against Mba and captured Bongo. Within hours, French paratroopers had landed at Libreville airport, liberated Bongo, and put Mba back in power.[29] The operation was led by a former French secret agent[30] who had founded the oil company that became Elf.[31]

President Mba, by now paranoid and ill, tried to resign but was not allowed to. An anonymous tract was circulated, revealing his secret traditional name, Mavego-Ma-Mididi: the Cat who can be Soft and Fierce at the same time.[32] In this superstitious land the exposure of his secret was a savage blow. Mba became a fearful recluse, surrounded by French advisors who got him to open the door further to French companies, hungry for uranium, timber, manganese, and, of course, oil.[33] "Under Mba, Gabon was so much under the influence of France," wrote one French author, "there was a risk that other countries would not realize it was truly independent."[34]

One day in 1965, Mba told the 30-year-old Bongo to go to Paris to meet President de Gaulle. "For a young African, to be received one-on-one by General de Gaulle: that was quite something," Bongo remembers. De Gaulle quizzed him about everything—what he thought about the French, about politics, and about French international relations. "The country north of Gabon, is . . . ?" de Gaulle trailed off, pretending to forget. "Cameroon, Mr. President," Bongo replied, starting to see that this was some kind of test.[35] The next year, Bongo was anointed vice president, by which time Mba was in a Paris hospital dying of (what is officially recorded as) cancer. Mba died in 1967 and Bongo, at just 32, became the world's youngest president.

The day after I landed in Libreville in 1997, the mysterious Mr. Autogue flew in from Paris first class on Air France, with a middle-aged female assistant with viciously plucked eyebrows. He was in his late forties, with khaki shorts, knobby white knees, and little rectangular-framed tinted glasses. He met me by a jetty near my hotel, where a boat was waiting to take us across the estuary to Coco Loco, a paradise beach restaurant among coconut trees, where white people jet-skied off the beach and buzzed overhead in microlight aircraft. I hadn't wanted to go, since I was tired, but he insisted. At Coco Loco he introduced me to a dazzling dirty-blonde woman in a bikini, who said she was an Air France stewardess. She sat beside me on a deck chair sipping cocktails, politely attentive. In front of us a muscled fisherman in blue trunks hacked open a five-foot barracuda. "Voilá!" Mr. Autogue said with a sweeping hand gesture across this playground. "Libreville!"

A Coco Loco employee came to chat. She regaled Mr. Autogue with the latest colorful corruption gossip from town. He gestured at her as a teacher would to a rowdy class: eyes wide, both hands, palms down, rising and falling. Shhhhh, shhhhh! We ate fish, the stewardess wandered off, and Alain got down to business. I should clear my calendar, he said. He would set up any interview I wanted—anyone except Bongo. He tried to get me to move into the Meridien Re'Ndama, a revoltingly expensive marble-lined hotel on the seafront, with expensive African art and jazz musicians. I resisted that, even after he suggested unspecified help with the bill. On a tight budget, and wondering what bugs might lurk in the Meridien's rooms, I stayed put at my own cheaper hotel, despite the voluble racist opinions of its French manageress.

The beautiful stewardess, still in bikini and flip-flops, joined us for the trip back from Coco Loco, but I opted not to talk to her. She stared out to sea and said goodbye when we landed. She was really something. From time to time I still wonder about her.

When I had planned this trip, I expected to struggle to get access to the politicians. I remembered long, humiliating, pointless ordeals in Angola trying to get interviews with ministers; and besides, my surname Shaxson sounds a bit like the dreaded words "Anglo-Saxon"—surely a disadvantage in Libreville, where it might be a bit like being called Shaddam and trying to poke around politics in Washington. Yet to my astonishment, it was effortless. Mr. Autogue pressed a cell phone into my hands (which I should probably not have accepted) and the appointments—every interview I had requested—slotted into place. The oil minister,[36] who was President Bongo's son-in-law? No problem. The finance minister? We shall have lunch with him. I enjoyed a long chat over cocktails at the Meridien with Jean Ping, the powerful half-Chinese foreign minister (later to become president of the United Nations General Assembly), who even guffawed at one of my bad jokes. Ministers received me exactly on time, and deferential secretaries had been expecting me. It was bizarre, like surfing a wave. In this land of fast cars, tropical spies, and gold trinkets, everyone I met was rich, and was grinning at me like Cheshire cats. Mr. Autogue clearly wanted to keep undesirables away; I did not mind much since I would have another week here after he left (when I did, the opposition parties treated me with suspicion; I felt that they suspected me of being a British agent). But while Libreville seemed prosperous, it just seemed dull and expensive. I felt like Alice in an African Wonderland. What else might be down this rabbit hole?

After becoming president, Bongo set about building a one-party state. "Gabon was being tugged between different tribes and regions," he said. "Having multiple parties would have been dangerous: democracy soon turns into an explosion of demands which it is impossible to satisfy, leading to endless strikes which fragile countries like ours cannot withstand." Gabon, he said, needed discipline, not democracy.[37] So he held an election, winning with over 99 percent of the vote.

Bongo remained fiercely pro-French, resisting advice from President Mobutu of Zaire to rename Gabon's second city, Franceville, with a more African name. "I love France, which is still my own country, a little bit,"[38] Bongo said. Yet he was becoming more confident and independent-minded, too, and he took Gabon into OPEC in 1973 and converted to Islam, changing his name from Albert-Bernard to El Hadj Omar Bongo.

Perhaps one of Bongo's greatest pieces of luck, oddly enough, was to have come from a minority ethnic group. This made him perfect for France: since he had no solid local support base of his own, he would have to rely heavily on French soldiers to prevent coups. But it also meant that he has had to work tirelessly to keep potentially antagonistic factions happy: building coalitions among minorities as counterweights to the dominant Fangs, and allocating resources judiciously to keep everyone loyal. Bongo began to prove himself a master puppeteer, deftly tweaking the strings of a helter-skelter of marionettes beneath him—from the powerful

"Bongo barons" plugged into leaks in the budget, through foreign politicians and oil, timber, and manganese interests, to the various leaders in a complex and antsy ethnic patchwork. "To fight tribalism, we made sure never to favor one of our nine provinces over the others," he went on. "If there are nine places to fill, we take one from each province. Really, it is a policy of quotas: we push minorities forward; a bit like what the Americans call positive discrimination. We try to choose the best person for the job, but it is not always possible. If the minorities feel aggrieved, they react, and it can be violent. Once a fabric is torn, it is hard to put it back together." He rebukes ministers who stuff their cabinets only with their ethnic brethren. Local politics are so sensitive to ethnicity that Bongo, even if provoked by interviewers, will not mention the ethnic groups by name. He also acts like a village superchief, personally resolving disputes. When students or trade unions clamor to go on strike, only a face-to-face interview with Bongo, where he might hand out cash, will solve it. When he is out of town, nobody in Gabon makes important decisions. "In your country," Bongo told a French interviewer, "numerous barriers prevent ordinary citizens from reaching the president. Not here. I receive ordinary citizens . . . some have come hundreds of kilometers to tell me their problems; I don't have the right not to receive them. Often I just give them what I can, a financial donation."[39]

In essence, he learned how to allocate oil money carefully. To outsiders, this looks like corruption—which is also what it is. Economics is about who gets what, so economics in Africa's oil zones is, in fact, just politics. He doesn't see it as sleaze. "Don't speak to me about corruption," he told one interviewer. "That is not an African word."[40] Yet he and his family also became—especially after the 1970s oil boom—fabulously wealthy. The French spy Maurice Robert, who became France's ambassador in Libreville in 1979, said that for all his admiration for Bongo, he had to try and protect Bongo against himself. "Spontaneously generous, he gave too much leeway to his family and close allies in the world of business," Robert said. "This stained his image badly, with accusations of corruption."[41]

Along with his extensive secret services at home and abroad, which mingled with the French foreign intelligence networks, other very curious currents underpinned Bongo's rule and the relationship with France. Secret societies like *Bwiti*, which use the drug iboga to conjure up spirits, are central to Gabonese culture. They use cruel, unpleasant, and even dangerous initiation rituals, which instill fierce loyalty. The writer Pierre Péan describes initiation rites for one secret society:[42] would-be ministers would travel to Franceville, near Bongo's birthplace, where they would make secret vows. Next, Bongo would wash his feet in a bowl, and the supplicant would drink the contents.

The French secret services knew that white people would not be initiated into these very African, very private power networks. But they did find that western fraternal societies—Freemasons, Rose-Croix (Rosicrucians), or Opus Dei—which feature prominently in the recent blockbuster book *The Da Vinci Code* and which still infest western politics today[43]—grew splendidly in Africa's fertile cultural soils, where local traditions had already learned over centuries to accommodate Catholicism, Protestantism, and other foreign beliefs. Bongo created (and is still today grand master of) the Grand Rite Équatorial,[44] an Africanized hybrid of two French lodges that has members in several countries, and most top Elf officials were masons. In the former British colonies, freemasonry is also active, but most are shunned by the Franco-African masons as Trojan horses for dreaded Anglo-Saxon influence.[45] A Gabonese professor told me more[46]:

> Freemasonry is about solidarity: you help others. It is also a way to control people, and those in power use it. These people are hard to recognize. Bongo uses freemasonry; he uses Rose-Croix; he uses *Bwiti*. With rituals there is fear of the sacred, of mystical forces. They can make you do things to use against you later. . . . If they have your secrets, you will obey. If you have the top 200 men in [this] position—then you have the country! There are other fears, too. You can prevent a man from earning: if you are in the opposition, you take Bongo's money. If you are cast out from the system that is the end for you: social death. That fear is as effective as violence. In some countries the fear of a brutal ruler is open, but not here. It is not like the death squads in Colombia. It is more subtle.

These fraternal societies supercharged Foccart's networks, which spread under, and subverted, the facades of official power, like tree roots. Fortified with oil money, they helped Bongo bend the leaders of fractious tribes to his will and give him a purchase on power at all levels of society. When the formal laws that we are familiar with in the West break down in Africa, solidarity networks replace them: the only order available amid the disorder of venal, broken nations. (Networks of different kinds also enable certain groups—Lebanese, Indian, Chinese, or Muslim traders—to flourish amid Africa's chaos.) By plugging Gabon into France via oil and freemasonry, Foccart had created something unusually potent. These systems also provided African rulers with side channels into French politics, linking up with what the writer Stephen Smith calls "a parallel, shadow structure, more important than the visible, elected structure of the French state."[47]

The invisible oil-fed subterranean networks help explain odd things. Opposition parties never get anywhere: they appear, but instead of falling under bullets or truncheon blows they rise, crest, then sink into irrelevance, as if sapped by a mysterious undertow. One firebrand opposition leader, Paul Mba Abessole, who roused the masses against Bongo in the early 1990s, is now part of the "presidential majority," a group of wealthy pro-Bongo political parties that one satirical Gabonese newspaper called "a box of cheese, assaulted by mice."[48] Others, informally labeled the "convivial opposition," pretend to oppose Bongo, but don't. These politicians drive luxury four-wheel drives and wear the latest suits from Paris. The late French author François-Xavier Verschave told me that when he was being sued jointly by Bongo and by the presidents of nearby Congo-Brazzaville and Chad, ordinary citizens from two of the countries lined up to defend him. But from Gabon, nobody came. "Gabon," he said, "is where the fear runs deepest. If you can instill that kind of fear, you don't have to do much of the rough stuff."

Foccart also created a covert strong-arm brigade, the Service d'Action Civique (SAC). This was first set up to check communist party trouble-makers at political demonstrations organized by president de Gaulle, but it evolved to carry out messier operations against France's enemies and to give Foccart's networks a steel backbone as he coiled his influence deep into Africa. The mercenary Bob Denard, the most famous of these men, oper-ated out of Libreville. With the help of French secret agents, Freemasons, oil, timber, uranium and manganese magnates, and the mercenaries and hard men, Bongo set up what has been called the Clan Gabonais, a group of operators loyal to French political parties, to France, and to Bongo. His Garde Presidentielle was a praetorian guard under French commanders, with French and Gabonese soldiers.

Gabon's oil was central to the system. The country became, according to the writer Pierre Péan, "an outgrowth of the French state; an extreme case of neocolonialism, verging on the caricature. . . . The roles are split, under the guise of noninterference. . . . Africans deal with tribal quarrels while the French look after serious things, namely exploiting the natural riches." Gabon was ruled jointly, he said, by Foccart, the right-wing Gaullist party in France, and Elf.

Years later, French investigators in the Elf headquarters in Paris seized a plan, never executed, to assassinate the writer Péan, who had nosed too far into the secret Franco-Gabonese world.[49] The book had been published not long after the election in 1981 of François Mitterrand, France's first Socialist president in a quarter of a century, and tremendous underground political tussles were underway between left-wing and right-wing factions in France, with Bongo leading the charge on behalf of the right-wingers. One incoming French foreign intelligence chief was astonished to discover that Elf's intelligence services were penetrating his own.[50] "Péan's book was interpreted by the Foccart people as a Socialist-inspired attempt to discredit them," said the Africa expert Stephen Ellis. "[The rise of Mitterrand] provoked real trepidation among the Elf networks, which feared the Socialists would destroy them. There were some on the left of the Socialist Party who indeed intended just that. A lengthy political struggle ensued that finished with an almost complete victory by the old Foccart networks, whose effective leader was Bongo." As part of his strategy, Bongo even threatened Mitterrand with the nuclear option: turning toward the Americans under the recently elected Ronald Reagan.[51] But he never did. Instead, a compromise was reached, which left the Clan unscathed: French right-wingers kept their fingers firmly in the fabulous Gabonese pie, but it was expanded to let the French Socialists get a piece too. It is remarkable to think how far this African president had influenced politics in a big western democracy.

Libreville also became a rear base for Igbo secessionists in Nigeria's Biafra war, whose cause de Gaulle had foolishly backed, hoping to split Nigeria's oil zones off from the English-speaking parent. About a million people died in that conflict. Over the years the Gabonese Clan squeezed its tentacles into conflicts in Rhodesia, Angola, Benin, Yemen, the Comoros Islands, Zaire's Katanga province, and Congo-Brazzaville; Bongo even recognized apartheid South Africa's racially segregated Bantustans. Occasionally, when it suited France, the United States used Bongo for its own international parallel diplomacy, such as paying off separatist rebels in the Angolan oil enclave of Cabinda where American oil companies operated. Bongo turned up in the most unlikely places: mediating in wars, calling African presidents, discreetly flying opposition figures or rebel leaders to Libreville for chats. They sometimes left with suitcases of cash. "The Gabonese miracle would give Gaullist France the means to achieve its ambitions, far from the concessions that it had to negotiate with the Anglo-Saxons in the reserved lands of the Middle East," wrote the French newspaper *Le monde diplomatique*. "From its Gabonese platform, Elf and its clan played politics across the African continent."[52]

Though Bongo was less repressive than many African leaders, his government was no soft touch. When a radical opposition leader[53] was shot dead in his car in Libreville in 1971, his family saw muscular white men pulling his body into a Peugeot and driving off.[54] The murder was never solved. In 1990, the year that Mitterrand made his famous "La Baule" speech supposedly marking a break with a dirty Franco-African past, a leading Gabonese dissident, Joseph Rendjambé, was found dead in his hotel room with syringe marks in his stomach. An inquest said he died of natural causes,[55] but disbelieving mobs, in Gabon's worst riots ever, briefly occupied the French embassy and took French oil staff hostage. French troops in armored vehicles quelled the riots, killing five. Not long afterward, Gabon hosted a huge tender for oil licenses.[56] Dozens of firms from Europe and North America prepared their dossiers, and submitted their bids. Elf won every license.

Not only did Elf get the best exploration territories and fabulous contract terms, but other French companies more or less kept up the monopolies they had enjoyed for over a century. Beyond all this, France also wrested control of much of the money from Gabon's natural resources that had escaped Elf and flowed contractually to Gabon's treasury. After independence, the central bank[57] had its headquarters in Paris, under a French chairman and a French-dominated board. The French treasury guaranteed the CFA franc currency, which Gabon shared with its near neighbors and which bolstered French control over its former colonies like Gabon. They had to deposit two-thirds of their reserves with the French treasury and to prepare annual programs of imports from nonfranc countries; France not only set limits on imports of some items from outside the franc zone, but also *minimum quantities* for imports from France.[58] Capital flight from Gabon financed French deficits.[59]

France had also set its relationships with its former colonies in stone with the now-infamous Cooperation Accords,[60] a kind of all-conquering version of foreign aid. A French prime minister, just before the wave of independence in 1960, explained how the colonies were to kowtow to French will, even after decolonization[61]: "Independence is granted on condition that the State, once independent, will abide by the cooperation agreements . . . one does not go without the other." These accords incorporated the secret defense accords, anchored the monetary and trading relationships, and secured strategic minerals for France. Some "aid" money flowed out of the French treasury, into the Gabonese offshore turntable, then back into France to secretly finance its political parties.

The Clan grew fat on a Franco-Gabonese bureaucracy of nested rules, tax breaks, and exemptions that were hard for outsiders to penetrate. Ports were awkward for ships not owned by the French transport and construc- tions titans[62]; imports without letters of credit from French banks faced un- predictable delays; and French weapons, railways, or airplanes could be serviced only by French engineers with French spare parts. From the French firms, Foccart harvested a bounty of intelligence; at the ports, air- ports, and finance ministries of the CFA Zone, nothing important escaped his notice. Even in 1997, when Mr. Autogue was presenting me to Gabon's captains of industry, some had white Frenchmen at their shoulders, politely massaging the interviews.[63]

Gabon also served French exporters a bonanza: for decades after inde- pendence, typically two-thirds of Gabon's imports have flowed from France, compared to just a hundredth from its African neighbors.[64] In a big supermarket in the oil town of Port-Gentil, I found no African produce *at all.* Instead, the shelves creaked with French milk from Choisy-le-Roi, French quail terrine, French duck confit, French televisions and videos, and racks of French wine. Even a frozen chicken that had grown up in Brazil found its way to this supermarket via Paris. Monopolistic practices boosted the profits of French companies in Gabon. The result, as the oil dollars also raised demand in the economy, was higher price levels, which encouraged more imports, while local producers suffered. It is no coincidence that at the turn of the millennium Libreville was the world's fifth most expensive place to live after Tokyo, Osaka, Hong Kong, and Zurich.[65]

Bongo became a flamboyant big spender. At an Organization of African Unity summit in Libreville in 1977, during the oil boom, half the national budget was spent on impressing visiting presidents, who got an honor guard with red velvet capes and gold swords. The skeletons of their hotels and armored Cadillacs litter Libreville today. Bongo also kicked off the Transgabonais rail project, to connect his home region in eastern Gabon with the coast. Launching it on his birthday with dancing girls, French politicians, and champagne, he promised a zone of prosperity one hundred kilometers wide along the railway's axis.[66] "If one must strike a bargain with the devil to build the railway," he said, "I am fully ready."[67] Today a surprisingly punctual air-conditioned rail service runs through the forest, and conductors in French-style peaked caps clip your tickets, while waiters serve croissants. Yet at a cost of almost three billion dollars, it was a sinkhole for the oil money; in a privatization tender in 1999, Gabon sold a 20-year concession on the railway, along with the rolling stock, to French timber interests who paid less than a hundredth of what Gabon had originally spent to build it.[68] If you take interest into account, its final cost is greater than today's national debt, most of which is owed to France.[69] Debt repayments today take up almost half of the budget.[70]

The late François-Xavier Verschave entitled one of his books *La Françafrique,*[71] referring to what he calls the "submerged face of the Franco-African iceberg." *Françafrique* is a play on the French words for France-Africa, but it also sounds like *France à fric,* which means France on the take.

We should not forget, however, that this system has also helped preserve Gabon for decades as an island of stability in a turbulent region. Some argue that this alone is enough to justify all the predation. Indeed, French scholars talked about a "pax franca" all across Francophone Africa. A confidential French military study in 1991 tried to appraise how the French, British, and other postcolonial systems had fared during the Cold War, producing results for the "grand experiment" that the Ivorian president Houphouët-Boigny had posed more than three decades earlier. It concluded that military spending had been proportionally lower in the former French colonies than in the British, Portuguese, or Belgians ones, but only 40,000 people had died in Francophone wars, compared to 2 million in the former British colonies, 2 million in the Belgian ones, and 1.2 million in the Portuguese ex-colonies.[72] The study was certainly biased: it ignored a million deaths in Biafra—many, if not most, of which resulted directly from French interference. Even so, it gave the French grounds for feeling smug.

The Franco-Gabonese Clan prospered through the turbulence of the end of the Cold War, and a little way beyond. But in 1994, as a new world order began to emerge, a decades-long political consensus in France about Africa suddenly began to break down. Change was in the air. Soon, Houphouët-Boigny's "grand experiment" to judge the relative strengths of the French and the other European models of decolonization would require a different appraisal, rather less kind to France.

EVA JOLY

ELF AFRICAINE AND THE RABBIT WARREN

The Alice-in-Wonderland domain that the mysterious Frenchman Alain Autogue had first introduced me to in 1997, I slowly discovered, was just a small antechamber in a very busy, and much bigger, secret rabbit warren, which was now beginning to come to light.

In 1994, the chief executive of U.S.-based Fairchild Corporation had filed a lawsuit in a commercial dispute against a French industrialist, Maurice Bidermann, triggering a French stock exchange inquiry. The case was referred to a prosecutor, then to an investigating magistrate, the Norwegian-born Eva Joly.

This was a year when the Africa policies that France had been pursuing since World War II suffered tremendous shocks. Under IMF pressure, the West African CFA franc, which had been locked for decades at a fiftieth of a French franc under a French treasury guarantee, was devalued overnight, to a hundred to one. Enraged African elites, who had not been pre-warned, saw the real value of local assets cut in half by this stroke of a Parisian pen, and the local price of champagne, luxury cars, and foreign debt doubled overnight. Riots erupted in west and central Africa. It was also the year of the Rwandan genocide, when French troops tried to oppose English-speaking Tutsi rebels in their sphere of influence, and in so doing ended up supporting the *genocidaires*, who killed hundreds of thousands of Rwandans with machetes and burned them in churches. This ghastly error dealt France's old Africa policy a body blow, and as Eva Joly's investigations began, the whole France-Africa edifice—La Françafrique—was trembling.

At first, her investigation into Bidermann generated just a few paragraphs in a few newspapers. Nobody expected her to puncture the impunity of France's elites. "But in my drawer, a fire was smoldering,"[1] Joly later wrote. Over the next eight years, the affair was to become the biggest fraud investigation in Europe since World War II. As she got closer to the heart of this matter—and the strange world of Omar Bongo's Gabon—the threats mounted. She was soon to fear for her life.

Eva Joly, born Gro Eva Farseth, grew up in Norway, the descendent of farmers and shipbuilders. Her childhood, almost in the arctic circle, was by her account a time of simplicity and innocence. In the summers her father, who made military uniforms, would take her, her mother, and her two sisters boating up the coast; they stopped for beach picnics and slept in log cabins. "Nature envelops you there," she wrote. "You live in osmosis with it." She spent much of her childhood at her grandmother's home, taking first a boat across the Oslo fjord, then the tram, to a school where she suffered hours of austere Protestant religious ideology; this was more than she enjoyed, and she blames it for distancing her from her emotions for years. As a teenager she served in a café and sold chocolates in a cinema. "If anyone had asked, I would probably have said I would get married, work in an office, live in an Oslo suburb, and holiday on the islands."

At university she chose philosophy, Latin—and French, which meant working overseas for six months. In 1964, aged 20, she went to Paris to work as a nanny in an upper-class family, the Jolys. She loved Paris, but the family was different. "The borders were sharply defined," she wrote. "I was never invited to the family table. The fresh bread was for them; yesterday's bread was for me." She was reprimanded by the mother, an elegant Paris sophisticate, for allowing one child to get mud on his clothes, after a stroll in the Bois de Boulogne. "For a Norwegian," Joly wrote, "this was the behavior of Chinese mandarin."[2]

Yet she was just a year younger than their son—and the two fell in love. This was a shock to the refined Parisian parents: her French was beneath society standards, and her name "Gro"—which sounds like *gros*, the French word for fat—did not help. "You must not marry this girl," Mr. Joly told his son. "Gro is not rich and is unlikely to become so. We don't know her family. Think how coarse her features are."[3]

Yet the couple married, and the Jolys severed their son's allowance. Eva Joly set about steeping herself in France; she even took courses in pronunciation to eliminate her accent (she failed). Gro and her mother-in-law slowly reconciled, and Gro took her advice and changed her day-to-day name to Eva. The couple had children, she studied law, and she passed the exams at age 37. "A country is made up of one sedimentary layer upon another: you can read its history in its laws," she wrote. "I learned French society from the inside."

She took a job as a legal advisor at a psychiatric hospital, then moved to a criminal tribunal in the city of Orléans. "Bonjour Monsieur!" she greeted the prosecutor on her first day. He looked her up and down, with disdain. "Madam," he replied, "for the period we shall spend with each other—which I hope is short—please call me Mister Prosecutor. It is not as if we used to look after pigs together!"[4] Joly set about mastering the protocol and the pleasantries, and slowly inserted herself into the rhythm of the place. Being a foreigner was not all bad, as she was unclassified in the elaborate French social hierarchies, so she often got away with being clumsy with French mores and taboos.

Yet for all France's byzantine social conventions, meritocracy in politics is alive too, and Eva Joly—fast, effective, and smart—raced up the ladder. She moved to the finance ministry, becoming the first female deputy secretary-general of a major department,[5] then, in 1993, she joined the Financial Brigade, a tribunal for investigating fraud.

"I don't see the point of these huge cases which last forever," the president of the tribunal told her when she arrived. "I bet you chose the financial cases because you are Norwegian and Protestant."[6] The subtext was that Protestants were interested in money, but Catholics did not want to dirty their hands with it. Some also say that while Protestant Anglo-Saxons distinguish between good money and dirty money, the Catholic view, pervasive in France, is that all money is tainted by sin, so is to be treated with suspicion.

Her offices lay in a dingy far corner of the Palace of Justice, which got unbearably hot in summer because the fumes from police cars milling outside meant they could not open the window. They had antique telephones, no fax or computer; just an Olivetti electric typewriter that clattered annoyingly. She borrowed her daughter's computer, and spent her own money on a fax. "Our work," she said, "didn't interest anybody."[7]

France's judicial system, for all its faults, has great strengths, too. The American and British justice systems are adversarial: defense and prosecution joust, and a jury decides on a defendant's guilt. The court is an impartial referee.

This is efficient but can be unfair; rich parties with better lawyers have an advantage and can reach deals to stop their dirty laundry being aired. French investigating magistrates, neither prosecution nor defense, are more like detectives; they follow their noses in a quest for the truth, and it is theoretically hard to stop them. When the system was created in 1808, Napoleon called the *juge* the most powerful man in France.[8]

But the French system also has mechanisms to protect powerful people. When a crime is suspected, a prosecutor must agree to the investigation. Although investigating magistrates may be independent, the prosecutor, whose career depends on the minister of justice, is not. Investigations can be neutered using, for example, *saucissonnage:* chopping cases into pieces like a sausage and assigning them to different magistrates who will then lose the plot. Also, if you are convicted and you appeal, you are not guilty until your appeal fails. Members of parliament who have been sentenced to jail keep working, and their cases often end up drowning under a backlog of tens of thousands of appeals. What is more, it is not possible to prosecute a standing president. "In England you protect the judges and the parliament from the executive," Joly said. "But in France you protect the executive from the judges. It is a tremendous cultural difference."[9]

Joly found that Elf Gabon, the local subsidiary of the French state oil firm Elf Aquitaine, had been funneling cash through Switzerland and Luxembourg to the industrialist Maurice Bidermann (whom she was probing),[10] to pay for a divorce settlement for the head of Elf, one of Bidermann's friends. She had opened up a paper trail that would reveal Elf Gabon—also part-owned by the Gabonese state and by Bongo—as the origin of giant, oily offshore slush funds for funneling cash to French political parties, and for tides of corrupt money sluicing around the globe.

Joly faced an intimidating old-boy network, a caste of mandarins from elite finishing schools like the École Nationale d'Administration, who have controlled political power and the bureaucracy for centuries, rotating among political posts, the secret services, and state companies like Air France, Thomson-CSF, and, of course, Elf. The system of revolving doors is sometimes known as *pantouflage*—from the French word for slippers—because moving from one post to the next is so easy that when you get to your next job, your slippers are waiting for you. The system produces politicians with brilliant minds, but it is also an exclusive club whose members look out for each other. Corruption—the abuse of public office for private gain or the misuse of public power—is a crime of the elites, and French high society considered itself untouchable.

"These civil servants may well be France's best and brightest, but they are also its most arrogant, a law unto themselves," wrote the American commentator David Ignatius.[11] "Over the years, their power has become subtly intertwined with a darker side of the French elite—the tough guys from Corsica and southern France who hold key positions in the police, the intelligence services, and the security departments of major French companies."

None was more intimidating than Charles Pasqua, a star in the Franco-African galaxy[12] and a former president of the right-wing band of hard-man enforcers, the Service d'Action Civique (SAC), which had used Gabon as a base for armed operations around Africa. Pasqua loved to spook his visitors. "Ah, my dear friend," he would say at the start of an interview; "it has been a long time since I looked over your file."[13] Helped by an anti-American populism, offshore African backing, and powerful clan networks from Corsica, this old friend of Chirac's had risen to become French interior minister. Corsicans figured prominently in the colonial service long before independence, no doubt because many of them wanted to escape the poverty of their homeland. After independence there were lots of them in the Franco-Gabonese clan. You can often spot Corsicans because their names, while French, sound Italian: Marchiani, Tarallo, Tomi, Leandri, Guélfi, Feliciaggi. "These mandarin . . . strands of power had become tightly intertwined in a network that has been dubbed 'France Inc.,'" wrote Ignatius. "The ruling clans needed each other—and they protected each other."[14]

Normally Bidermann's case would have been quietly dropped. But Joly was an outsider, and French politics was in flux after the fall of the Berlin Wall. The French left and right, as well as the foreign and domestic secret services, were engaged in giant, tentacular struggles, which were generating press leaks and anonymous tip-offs for the magistrates. Elf was being privatized then, too, and the new head of the company lodged a formal complaint with the magistrates against his predecessor, hoping to mark a clean break with Elf's dirty past.[15] French president Jacques Chirac, who took power in 1995, also wanted to tarnish the image of his predecessor, François Mitterrand, and his people were adding to the leaks. These people saw Joly as another weak judicial genie—a foreign one at that—who could be coaxed or bullied back into her bottle later.

Yet Joly had already shown her mettle by jailing Bernard Tapie, a crooked former minister and president of the Olympique de Marseille football club. And, as she picked up the Elf affair (alone, until another magistrate, Laurence Vichnievsky, was assigned to help her in 1997), she enchanted the public, at first. This protected her.

The initial probe into Bidermann was like pulling on a rope sticking out from a swamp. As she kept pulling the rope branched, then divided again, and again, and she began to drag exotic political creatures, spluttering and protesting, out from the turbulent, murky swamp of French politics and into the sunshine.

At first, the mandarins seemed stunned. Then, slowly, they began to marshal their response.

A key to Elf's subterfuges was to split the paper trail between countries. There is no global police force navigating the fragmented world of international finance, putting the pieces together. Joly followed financial transactions to the French border, where her investigations hit a wall. To get foreign cooperation, she had to send her requests up through the French judicial hierarchy, where they might stew for months, before being transmitted grudgingly overseas, then often regurgitated back for niggly procedural reasons. Often, appeals were launched against even the transmission of evidence overseas, delaying things further. "The judicial borders . . . we can't cross them. We need international judicial help to do that—it's a very old-fashioned system and it's not working well."[16]

Countries are desperate to attract foreign capital; one way to do this is by allowing their most illustrious banks to shroud their clients' money in secrecy, which helps attract this kind of dirty money. "Laws are made inside states. Money flows across borders," Joly wrote. "For international criminals, impunity is assured."[17] Liechtenstein, the Isle of Man, Gibraltar, the Cayman Islands, Bahrain, London, and so on: her investigations were stonewalled. But there was an exception: a few crime-busting Swiss magistrates who, like Joly and like brave Italian magistrates tackling the mafia, stood firm. This link with Geneva provided the key to some of Elf's mysteries.

Yet bank secrecy was not Joly's only obstacle. She began to get threats, even death threats, often from unexpected quarters. A friend introduced a man called Franz, who took her aside one evening. "Madam, 98 percent of felonies can be judged, but two percent cannot," he said. "These are the state secrets. There are many powerful interests around you. Beware: state secrets have their guardians, and they are not gentle."[18]

She was followed in the street. Pursuers stared at her brazenly. She got phone calls, with only silence at the other end. Once, neighbors saw different cars with identical license plates behaving oddly near her home.

Documents were stolen from her office, which was infested with spy bugs. She found a calling card stuck to her office door, with a list of French magistrates killed since World War II. The names were crossed out, bar one: Eva Joly.[19]

"I felt like I was penetrating an unknown world, with its own laws," she said. "But if they thought they would stop me with this kind of provocation, they were wrong."[20] She got armed police protection. The officers asked her to wear a bulletproof vest and they carried a Kevlar plate in a case, as extra protection; in traffic, they fingered their guns when motorcyclists approached. "No more meetings in cafés. No more window-shopping. The police would put their guns on my kitchen table and inspect all my mail before giving it to me." These were honest cops who took their jobs seriously, and Joly developed a good rapport with some of them. On Mother's Day, one bodyguard, who was built like a brick blockhouse, brought her flowers. "Madame," he said, "I would lay down my life to protect you." And she knew he meant it.[21]

She received a miniature coffin in the post; in one place she raided for evidence she found a loaded Smith & Wesson revolver, pointed at the entrance.[22] At an embassy cocktail party, an army general told her that if she started investigating arms, she would not last 48 hours. "I had assumed that people in general respected the laws," Joly later said. "But reality outstripped fiction. There was an ocean of fraud at the highest level."

It was essential to be meticulous about her investigations, for any slip would be exploited by her enemies. Lawyers set legal traps for her, and some leads were blocked by a legal mechanism called "secret defense"— which can be invoked to protect issues of "national security," stopping investigations cold.

Slowly, news organizations, and sections of a once-sympathetic French public, began to turn against her. "Part of the press turned. Or was it turned? After all, Joly and her colleagues were now accusing people who could pull strings at the highest level," remembers the British writer Tim King. "The press began by attacking her appearance. The photographs chosen for publication now showed a grumpy middle-aged harridan, worn down by 18-hour days. She was the 'former au pair,' the 'Viking,' and the 'Protestant.' A banker she was investigating mocked her accent and refused to answer questions unless they were 'in proper French.'"[23] She was imagining it all, some wrote; she had been watching too many Hollywood gangster movies; she was drunk with power, and she was mistreating honorable state officials. She was accused of being a Trotskyite, or a CIA agent sent to destroy French oil interests. She was a wily female manipulator seeking revenge for past social slights; she was pursuing a personal vendetta against the entire business world.[24] Journalists got dramatic leaked scoops from inside her investigations—then she was blamed for the leaks.

"Businessmen, some of whom had already been fingered for corruption, [had] moved their money into the media, knowing that no editor will publish defamatory material about one of the group's major shareholders," said a French policeman who had dug into political party financing in France (and was later sacked).[25] Joly herself talks of a "media-industrial complex"—reminiscent of President Eisenhower's warning in 1961 about the dangers of unwarranted influence from the "military-industrial" complex in the United States.

Meanwhile, her colleagues grew distant. "Not a word of sympathy . . . in the corridors, haughty indifference," she wrote. "I felt as if I were radioactive . . . it wounded me, but I saw that the hostility was directed not so much at me. It was their way of denying the reality of the Elf Affair."[26]

In 1997—the year that the enigmatic Mr. Autogue showed me round Libreville—she hauled in André Tarallo, the head of Elf Gabon. The pressure intensified. Tarallo was another right-wing Corsican Gaullist and a Freemason, who was in Jacques Chirac's class at the École Nationale d'Administration.[27] With a balding head crammed with secrets, and a silky sensitivity to African cultures where Elf operated, Tarallo was Elf's "Monsieur Afrique," arguably the most important Frenchman in Africa. The Socialists had never dared to oust Tarallo or to disrupt this cash cow for the French right wing, even after president François Mitterrand installed his ally, Loïk Le Floch-Prigent, at the head of Elf. Instead, France's political factions reached an arrangement, rather like Italy's *lotizzazione*, where political pluralism means agreeing to divide the pie among the elites.

"Le Floch knew that if he cut the financing networks to the RPR[28] [Chirac's party] and the secret services it would be war," wrote the journalists Valéry Lecasble and Airy Routier.[29] "He got this message from African presidents and the secret services. It was explained that instead, the leaders of the RPR—Jacques Chirac and Charles Pasqua—did not mind the Socialists taking part of the cake, if it were enlarged."

So the new left-wing head of Elf cooked up a new layer of dirty money, for the French left, leaving Tarallo in place to charm African presidents and to squeeze the oily udders of Elf's African cash cows. Le Floch-Prigent set up new networks with his own hatchet man, Alfred Sirven, another Freemason and former French Resistance fighter who dealt mainly with African opposition and rebel officials.[30] Sirven, who claimed to have enough information to "blow up the republic 20 times over," became famous for swallowing his telephone chip when captured in the Philippines, after Joly launched a global manhunt for him.

"Originally, everything was more or less Gaullist controlled," said the controversial late author François-Xavier Verschave. "Then Le Floch came. The system became more classical, more heterodox, more baroque. So it became vulnerable."[31] Events got so out of hand that Le Floch-Prigent once called the French foreign secret services, which were part of the system, "a great brothel, where nobody knows any more who is doing what."[32]

As the factions issued leaks and counterleaks about the other side, fresh information emerged. A central repository of secrets, the magistrates learned, was a drab-looking building called Les Frangipaniers in downtown Libreville next to the Elf offices, near the French and American embassies. This housed the French Intercontinental Bank for Africa (Fiba), a bank set up by a former governor of the Bank of France (another Freemason) in 1975, during the oil boom. Fiba's partners included Elf, Bongo, his children, and senior Gabonese officials as shareholders,[33] and it was overseen by Tarallo as the head of Elf Gabon. "If you do not understand Fiba," said Le Floch-Prigent, "you do not understand the Elf system."[34] Fiba also had offices in Paris and in neighboring Congo-Brazzaville, whose president Dénis Sassou Nguesso is Bongo's father-in-law. Several other African countries (or their leaders) opened secret accounts there, plugging other foreign oil industries into the offshore Elf system.

Like a trick bookcase in a haunted house, Fiba spun Elf's cash back to front and upside down, then snaked it out through tax havens.[35] Orders were transmitted verbally, then documentation was destroyed. African leaders got 20 to 60 cents of each barrel that Elf produced in their countries[36]; other flows went to the intelligence services for covert operations, a bit like the Iran-Contra affair in the 1980s (and the two affairs shared some participants).[37] Fiba's president said in court that financial flows were conceived so that the Africans were only aware of the official lending, but were ignorant of the whole system—which Elf rendered deliberately opaque.[38]

"People would know exactly how much money would fit into a suitcase," said the French Africa expert Stephen Smith. "You would be accompanied by an aide of the Gabonese president to Libreville airport so you would be checked through; you would get on the Swissair flight, get out in Geneva, deposit the money, then take the ongoing flight to Paris. That was routine."[39] (I met one of these bagmen once: he described transporting a suitcase of cash to an African rebel leader. I cannot give any more details of our confidential conversation, but I still marvel at this bagman's strange global political connections.) Bribes flowed through Fiba and Elf Gabon, as Elf dealt with politicians in Nigeria, Angola, Congo, Venezuela, Germany, and so on—not just for Elf, but also in pursuit of broader French diplomatic, commercial, and military goals. The crooked strands twisted out through the shadows of international finance, into Anglo-Saxon domains, to Spain, Guatemala, or Uzbekistan. Elf won contracts across the globe, infuriating the Americans, among others.

Elf Gabon's cash also paid for luxury apartments in Paris, free flights, mansions in Corsica, buying the silence of French and foreign politicians and so protecting the system. Elf channeled up to $120 million each year in official "commissions" (or bribes) for Elf's "normal" international business, and about the same again in "occult" payments, plenty of which flowed to French political parties.[40]

The magistrates dragged peculiar intermediaries into the ever-expanding nets: a Corsican nicknamed Dédé-la-Sardine (because he owned fishing boats) who was once the personal pilot of the head of the International Olympic Committee; the colorful late British billionaire Jimmy Goldsmith; and French, German, and former Soviet agents. Joly's biggest target was Roland Dumas, a former foreign minister and the president of France's Constitutional Court, who also had masonic links with Bongo.[41] With flowing white hair and a famous art collection, Dumas was a doyen of the Paris left, and newspapers gorged on lurid tales of his high-class mistress (who wrote a book about it, *Whore of the Republic*), $1,800 handmade boots that were supposed to be washed once a year in champagne, and bribery surrounding the sale of six military frigates to Taiwan. (Dumas was eventually cleared on appeal, but his mistress went to jail.)

While Dumas was in her line of fire, the pressure on Joly ratcheted up. "Sometimes an investigation is like a game of cat and mouse," she said. "In this one, I was the mouse." She had nightmares and dreamed she was being chased.[42] She worked day and night, carrying a portable fax to scenes she was investigating, to speed things up. She grew apart from her husband, and they separated. "The danger . . . it is harder for your family than for yourself: I have the action, I have the enquiries, so I have something to do," she said during the investigations. "It is so evident that the danger is there." At the height of it all, her husband, from whom she had grown apart, died. This is not something she has ever been happy talking about. For a time she fought on, but the battle took an increasing toll. "In the last months of the Elf affair, my resistance was hanging by a thread."[43]

In 2002 she threw in the towel, returning home to advise Norway's government on corruption, leaving other judges—independent people who entered the breach she opened—to finish the Elf case. They got their convictions in November 2003, the culmination of eight years of investigation. The former head of Elf and his hatchet man Alfred Sirven were each found guilty of misusing well over $100 million of company money and got five years in jail (where Sirven died); the right-wing head of Elf Gabon, André Tarallo, got four years, while 25 other underlings, intermediaries, and ex-wives received shorter jail sentences and fines.

The magistrates had exposed this dirty iceberg, but for legal reasons only its mucky tip, mostly the bits about personal enrichment, or misuse of the company's assets by private employees, were prosecuted successfully. The probes also focused heavily on just one short period—from 1989 to 1993, while Loïk Le Floch-Prigent was the head of Elf. He said what everyone knew: that he was just doing what all his predecessors had done, and that he was taking "daily" instructions from president Mitterrand. Elf, for its part, said afterward that it was a victim of embezzlement of company funds by senior executives acting mainly in their personal interests, and that it set up an ethics committee and training programs from February 1995 to avoid any recurrence.

The investigations could not tackle the wholesale bribery of African politicians, or Elf's payments to armed rebel movements, because bribery of foreign officials was not only legal, but tax deductible in France (this was the case then for many OECD countries).[44] Yet it was still a victory, of sorts. Joly's successors are following new leads today, in different but linked cases, and judicial cooperation between European countries has since become marginally better.

Now, Gabon is gradually diversifying away from France. Shell, not Total, operates Gabon's largest oilfield, and smaller, and nimbler, American, Canadian, South African, and—recently—Chinese companies, often better able to tease more oil out of declining fields, are moving in. Following its privatization in 1994, Elf was merged with TotalFina in 2000, and the new company, called Total, has a far more Anglo-Saxon flavor, and has, without a doubt, decisively broken with old ways. Total says the Elf system has been entirely dismantled.

But has corruption been dismantled in France? Who knows. The really big fish have eluded the courts. Notably President Jacques Chirac, who benefits from presidential immunity while in office and who dismisses the allegations against him as an "abracadabra tale." He was satirized in a French puppet show as Superliar—a latex hero with a briefcase full of cash—and is scorned by most of the French public today. But he, and others who might have fallen, remains in power.

"You wanted to change France, Madame!" a lawyer once told Joly. "But that is impossible!" And he erupted into laughter.[45]

Some things remains the same. While France has slashed its troop numbers in Africa, the 650-strong French military garrison in Libreville is still firmly in place, to protect Bongo's mansion. Though it is now supposed to be a base for regional peacekeeping efforts, it still powerfully deters coup plotters.[46] Bongo still visits Paris—less often than before—and he is still received by France's top politicians—and by a few non-French European ones, too.[47] These are not the courtesies normally afforded to leaders of little-known African countries with populations of less than two million people.

French foreign aid to Gabon appears to suggest a change: running at $100 million per year until the mid-1990s, the aid fell to $20 million in 1997, the year Gabon's oil production peaked, and the height of Joly's investigations. Aid flows even briefly turned negative in 2000, as repayments to France exceeded new loans.[48] You might expect aid to rise, not fall, as oil output falls and a country gets poorer. Perhaps the intermingling of aid money with other flows into the secret Franco-Gabonese system has now ended. Yet I still see odd things: "rural development" aid flowing to French timber interests, and $50 million in aid from the European Union (which has taken over part of French aid) to Gabon for mineral exploration from 2004 to 2010—a gift from Europe's taxpayers to French interests in Gabon. Perhaps this is just old-fashioned industrial support.[49]

Elf's successor, Total Gabon, still has Bongo's daughter Pascaline on its board of directors (she has the same role at Shell Gabon, too).[50] There have been "no particular repercussions" in Gabon from the Elf-TotalFina merger, a senior, unsmiling Gabonese official told me in an icy interview in Libreville in 2002. "As regards Total's closeness to the government of Gabon," he said, "not much has changed." A Total official I interviewed said that giving me information about investments, reserves, or even the share price, was "forbidden," and I must contact headquarters in Paris. It was "inconvenient" for him even to give me their phone number, or even a copy of the latest Elf Gabon report—a public document.

"The Elf Affair is the culmination of an Epoch," Joly wrote in 2003. "The signs are that the country will go no further, as if French democracy has reached the limit of the revelations it is prepared to tolerate."[51] The American commentator David Ignatius said it was as if Joly had netted a load of fish, which are now left to rot while the French public politely holds its nose. "She prized open a door, at great personal risk, but the political class refused to follow her through it into a new era of accountability. For a moment at least, the system was weak, exposed, and perhaps even ready to topple, but it survived because of the code of silence of the French elite."

Today the Elf affair tends to elicit yawns in France, not outrage. In 2006 the four top magistrates at the Financial Brigade, Joly's successors, were promoted elsewhere. "Financial justice is no longer a priority," the newspaper Le Figaro said, quoting the top departing crime fighter.[52] "The financial brigade in Paris appears to have been decapitated." This may be exaggerated, but there is little doubt that the bomb that was ticking at the heart of the French establishment seems to have been quietly defused.

The French public appears to have accepted, once again, what the French commentator Alain Minc has called "a normal Latin predisposition to deals and fiddles." Things could never get so bad in Britain or America.

Or could they? Eva Joly suggests an answer, singling out one tax haven she found particularly obstructive: "The City of London, that state within a state which has never transmitted even the smallest piece of usable evidence to a foreign magistrate."[53] Britain took three years to respond to an international arrest warrant issued by French magistrates in 2000 for Nadhmi Auchi, a former ally of Saddam Hussein and a big shareholder at the time in the BNP-Paribas bank, whom they wanted to question in connection with the Elf affair.[54] Germany did not get off lightly, either: another murky chapter involved the political party of the former German chancellor, Helmut Kohl.

Joly had been able to pursue her leads doggedly thanks to France's inquisitorial judicial system, which let her behave like an independent detective. In American justice, however, defendants can negotiate early settlements of their cases through plea bargains. This speeds up judicial processes but also can mean that defendants' fates often depend not on independent detective work but on negotiations between prosecutors and defendants. This gives rich defendants, who can buy the best legal advice, the advantage. Many cases that should come to court are prematurely ended because of plea bargains, and we never hear the real story.

In America there is one, very rarely used, independent investigative mechanism: the U.S. Senate Permanent Subcommittee on Investigations. One of its probes—into private banking and money laundering—looked into Citibank, the largest bank in the United States, which had accepted millions of dollars from the brother of former Mexican president Carlos Salinas. The investigations turned up other characters: the family of former Nigerian dictator Sani Abacha—and Omar Bongo. This was not a criminal case; just a Senate probe into how to make the U.S. banking system work better. It is perfectly legal—believe it or not—for U.S. banks to accept the proceeds of the looting of foreign treasuries by corrupt dictators.[55]

Starting in the 1970s oil boom, Citibank began taking tens of millions of dollars from Bongo, in a personal capacity. They set up shell corporations for him in Bahrain, Jersey, London, Luxembourg, New York, Paris, Switzerland, and the Bahamas, sending payment orders not through the bank's regular credit channels but only though certain senior bank staff. They kept the money out of Gabon's banks, so that locals could not leak information either. Citibank was supposed to ask about their clients' sources of wealth; but in practice they seem to have hardly cared. One $20 million transaction in 1997 involved a South African firm,[56] which Bongo had invited in to compete with French oil interests, partly out of pique at Joly's probes.

Bank documents, describing an "extremely profitable relationship," estimated Bongo's own wealth conservatively at just $200 million, derived from real estate and "many oil ventures" and his "connection to French oil companies [Elf]." Yet a Citibank official told the Senate that he never once asked Bongo about the source of his wealth "for reasons of etiquette and protocol." Another told the Office of the Comptroller of the Currency (OCC) in 1997 that Bongo would have a courier pick up suitcases full of cash from the oil companies, and he always paid cash when visiting the United States. On one visit to the United States, Citibank noted, Bongo's entourage took two full floors at the Plaza Hotel in New York.[57]

Bank officials go to great lengths to conceal the cash. An outside expert told the Senate how private bankers routinely posed as tourists when traveling overseas to meet their clients, partly because foreign governments often took a dim view of the politicians ransacking their treasuries. They converted bank account numbers into letters using secret codes.[58]

"Every year an overall allocation, loosely referred to as 'security' or 'political' funds, is voted into Gabon's budget," Citibank said in a memo. "Although not spelled out for obvious reasons, these funds are understood to be used at the discretion of the Presidency... without any limitation." Memos listed hidden funds totaling 8.5 percent of Gabon's budget. (Had they taken half an hour to check this with the IMF, they would have found that there is nothing remotely like this in Gabon's budget.)

For his part, Bongo admitted having his own money in the accounts; part of it, he said, was for lobbying in the United States, and the fuss caused by the Senate investigations, he alleged, was stirred up as part of larger political and economic struggles between France and the United States. Yet he had his own idiosyncratic explanation for how he spent the rest. His words, to western eyes, are tantamount to an admission of a kind of corruption. But they also reflect the realities of power in an oil state. To stay in power, you must keep paying off the restive factions to keep them quiet:

> I redistribute nearly all the money that the Gabonese state confers on me. The teachers don't have money, no cars? I pay. Demonstrators? I pay. Women's day? I pay. We have not reached your level of political maturity. Here, if people think the boss doesn't have money to redistribute, he isn't respected. Sharing is part of African society, at all levels. This money serves our democracy, which is not like yours. Our culture is different, but it is not inferior to yours. I don't know if this system is good. But it is the only possible way![59]

By 1997, as Joly's probes generated unhealthy news headlines, internal memos from the Citibank account managers were beginning to register unease about Bongo:

> All of us need to be very thoughtful and selective about the press coverage we choose to interpret and share.... We ought to be extremely careful about sharing such information with the regulatory authorities, because we can't answer for it. We should stay as far away as possible from this mess....
>
> P.S. Please take out a subscription to *Le Monde* in New York. Sabine, with respect, cannot be expected to scan and clip the local press on a daily basis.[60]

Then, by 1998, as the Senate probe got underway, new memos show Citibank officials getting even more jittery:

> We ought to face this issue and its implications with our eyes wide open. Both [...] and [...][61] [two bank officials] are already beyond Gabon.

They will not be around to pick up the pieces. . . . We should bear in mind that the relationship is deeply anchored in trust. It is a very personal matter involving much of his family vis-à-vis which he cannot afford to lose face. This is likely to magnify his reaction, with further blows to our credibility as a Private Bank. [Bongo] and Mandela are the foremost African leaders today, and they are friends. Although we can't measure how far the negative vibes could go, there is no doubt that they will spread and that will include France. [Bongo's] family and friends extend far. . . . There would be legitimate grounds for concern in many minds that Citibank was abandoning this part of the continent.[62]

Finally, Citibank closed the accounts. The press soon lost interest in yet another African corruption tale, and a Citigroup official painted it as a matter of a few rogue officials failing to implement proper controls. "I do think we are talking about five or six cases out of a large number . . . the private bank now has 100% good audits which was not the case during the period of time we are talking about."

Other experts called to testify about the general state of American banking did not quite see it this way. "American banks help wealthy customers do abroad what the customer and the bank can't do in the United States under U.S. law," said one expert, Raymond Baker, reeling off a list of foreign states—like nuclear-armed Pakistan and Russia, or Nigeria—where politicians use bank secrecy to hide hundreds of billions of dollars looted from their treasuries, with savage effects on their countries' political and economic stability. Corrupt money, Baker said, provides a cover for criminal money, estimated at as much as a trillion dollars per year, with half of it flowing to the United States. "Criminals no longer need to carry Samsonite suitcases of cash. Instead, they . . . take advantage of the products and services that private banks aggressively market.[63] Anti-money-laundering efforts are a failure. Illegal flight capital erodes U.S. strategic objectives in transitional economies. . . . The U.S. has become the largest repository of ill-gotten gains in the world."[64]

His words back up and echo Joly's conclusion of her shattering years with the Elf Affair. "What we see is not a singular phenomenon, is not a curiosity, is not individuals having lost their way," she said. "It looks like a system."[65]

The Elf Affair poses a question. Why did politicians use Elf as their slush fund, when offshore tax havens might have done the job, using public money as the cash cow? And why did they use Elf's African subsidiaries, not the parent company Elf-Aquitaine, as their black box? The answers help us see a bit more clearly why the oil industry operates outside the rules of normal free-market competition.

First, consider a "normal" company—one that makes, say, electric fans. It must stay on its toes, by constantly innovating and keeping costs down— or its competitors will crush it. Milk this company enough for corrupt purposes, and it will eventually fail.

Yet oil and gas companies are different. Oil provides "rents," as described earlier—money not earned by innovation and hard work but that comes out of the ground as if a gift from God. You can milk a company like this as much as you like, and "competitors" will not drive it out of the market because whatever it does with its profits, the oil is there under the earth, and will keep flowing. This can, especially when oil prices are high, generate giant balloons of cash, ideal for milking and putting to unholy uses. This kind of money is mischief money.

What is more, rich countries have systems and controls in place on behalf of shareholders, taxpayers, and democracy itself, to stem abuses. These systems are fallible, to be sure, and it is certainly possible to tease corrupt money out of other French companies in France, but this needs financial and political gymnastics to bypass the controls. There are limits to how far companies can be milked into tax havens before they collapse. With an off-shore oil company like Elf Gabon, the systems to control abuse are mostly not in the hands of French regulators, magistrates, and shareholders—but are in the hands of people like Omar Bongo. Having selected him because he is from an ethnic minority and has no natural support base, and therefore needs French soldiers for protection, France was well placed to ask him to make some of his balloon of oil cash available.

Some people have argued that the Elf system was good for France. Greasing foreign palms with Elf's money also helped French companies win lucrative contracts, and secured French strategic objectives around the world. Elf did well too. With Gabon as its springboard, it grew faster than its Anglo-Saxon rivals, and its successor Total has ended up as a beneficiary of past practices. Today, Total is the fourth largest private oil and gas firm on the planet—bigger than Chevron and 25 of the 30 companies in the Dow Jones Industrial Average.[66]

Yet—as always seems to be the case with oil—the real cost is bigger, and more nebulous. The cost has been this: nothing less than the subversion of French democracy, which is now in crisis. Rioters burn cars in the

Paris suburbs, as the political classes stumble from one scandal to another. A new affair over the clearing house Clearstream—also involving tax havens—is again troubling French politics. Massive public cynicism about the mainstream parties is feeding a rapid growth in far-right extremism. In France's 2002 two-round presidential election, the Socialist Lionel Jospin came third to Jacques Chirac (who won less than 20 percent, the lowest ever for an incumbent president) and to the far-right politician Jean-Marie Le Pen in the first round (35 percent voted for extreme-right or extreme-left candidates), leaving voters in the second round with a choice between Chirac and a far-right leader who presented himself as being free of corruption. "While these derisory polemics are amusing the gallery," Le Pen said, "the self-destruction of France is taking place on every front, with criminal indifference among those in charge."[67] Voters massively endorsed Chirac in 2002—just to stop Le Pen—yet many wore rubber gloves as they did so, to show their disgust. France also seems to have turned in on itself: in surveys fewer than 10 percent of French people see globalization as a positive force, and 85 percent of them think that their country is on the "wrong track."[68]

African oil did not create the system or its failings, and Elf's money also mingled with corrupt flows from domestic real estate deals and the like. But the African slush fund contributed powerfully to the corrosion of French democracy.

Could African oil be playing the same kind of havoc in other western countries? Similar poisonous political melds exist elsewhere, oiled with the same sort of mischief money. In his book *Sleeping with the Devil: How Washington Sold Our Soul for Saudi Crude,* the former CIA agent Robert Baer describes a bigger, dirtier, and even more threatening version of the offshore oil menace, which is corroding American democracy while also helping radicalize young Saudis and feeding al-Qaeda. This looks to me like America's very own Elf system—but there has been no Eva Joly in Washington to crack it open. Like in France, the American-Saudi meld is an open secret, infecting the entire political spectrum from left to right, and underpinned by a military relationship. Its critics, too, are often painted as paranoid conspiracy theorists. The American justice system might be failing the American public in probing oil corruption: as another example, current investigations into a $180 million possible bribe scandal involving the American oil services company Halliburton and a liquefied natural gas project in Nigeria was not broken open in the United States, but in Paris, by one of Joly's colleagues.[69] Britain, too, has its own strange Saudi relationship, again cloaked in secrecy and unchallenged by British courts, involving a giant, secretive arms-for-oil arrangement known as al-Yamamah, which has been described as "the biggest [UK] sale of anything to anyone."[70] Has it tainted British politics, or affected British foreign policy? It's hard to know; an investigation into it by the British National Audit Office (NAO) is the only NAO document never to have been published.[71] I have indications of one, and possibly two, more of these kinds of arrangements in Europe, linked to African oil, though I don't have enough information to go into them here. There are surely others.

To set up such symbiotic oil-based offshore arrangements it is not necessary for a western country to control an oil producing country, in the way France has controlled Gabon; such arrangements can be negotiated, too. The only requirement is to have a source of this mischief money, such as from oil. In the rich English-speaking world the corruption is probably not in the form of a grand unified French-style state conspiracy, but a more privatized, decentralized, nebulous rottenness, stemming from these grand, generic offshore temptations which insidiously poison our democracies through lobbying and other, more nefarious, schemes. The higher the oil price, the bigger the balloons of cash, and the greater the threat to our democracies. Once again, oil acts like heroin: it feels good, but the final result is catastrophic. This is not just more tedious foreign corruption. This is really dangerous stuff.

"I have seen so many resemblances, in France and abroad, between the corruption of the state and all the mafias," Eva Joly said. "The same networks, the same henchmen, the same banks, the same marble villas. . . . We think crime lurks in the shadows of our societies. But we find it linked intimately to our great companies and our most honorable politicians."[72]

Finally, did Gabon's oil help or hurt Gabon itself?

"Oil is just the best thing for this little country," the head of a small American oil exploration operation told me in Libreville. "They say to us, like, 'Hey! Where've you been?' They are just glad if we come and buy a few of their bananas!"

We know better, of course. But we should make a sober assessment, as the answer is not immediately clear. The oil-fed symbiosis between France and Gabon was certainly a shocking display of French neocolonial predation, and oil seems to have brought its usual malaises: life expectancy at birth in Gabon is just 54 years, seven years less than in nearby São Tomé e Principe,[73] whose GDP per capita is a fifth of Gabon's. One Gabonese child in seven is malnourished, and over 60 percent of Gabonese live below the poverty line.[74] For most of its post-1970s history, barring the latest oil boom, Gabon's economic growth was negative.[75] Before 1970, 40 percent of Gabonese lived in towns or cities. Today, three billion barrels of oil production later, over 80 percent do, the highest in the developing world after Djibouti, Libya, and Saudi Arabia. Agriculture is on its knees, and manufactured exports make up only a fiftieth of all foreign sales.[76] Gabon imports nearly two-thirds[77] of its food, and servicing its $4 billion of external debt, piled up in the good times, takes up a third of the budget. And so on. Now, with its oil running out, Omar Bongo Ondimba (he added "Ondimba" to his name in 2004) finds that the strings he used to pull in Paris to get out of economic crises no longer work so well. He has appointed his son (another Freemason) as defense minister; and many Gabonese think that his son is being groomed to succeed him.

Yet for all this woe, it is still not clear that French policy has been unambiguously bad for Gabon. Africa's ethnic cleavages certainly make independence and democracy potentially dangerous and destabilizing—and the French were right to argue that this was not to be embraced without forethought. French policy "has allowed us to come out better off than the former British, Spanish, or Portuguese colonies," Bongo said. "That is pretty much enough for us."[78] There is something in what he says: World Bank data shows Gabon performing modestly better in many aspects of human welfare than many other countries (especially oil-producing ones) in this sub-region.[79]

The real damage was done instead to French democracy and, as we shall see in the next chapter, to one of Gabon's neighbors mired in conflict: Congo-Brazzaville.

6

ANDRÉ MILONGO

GOLDEN EGGS

A while after I stepped through the looking glass in Gabon, an editor passed on a plea from the United Nations chief in the Congo Republic. He wanted foreign journalists—anyone, please!—to come and cover a humanitarian disaster unfolding there.

Aid officials there constantly lament that foreign donors and news editors overlook this Congo, as they focus instead on its larger, similarly named southern neighbor, the giant, war-stricken former Belgian colony called the Democratic Republic of Congo (DRC, previously Zaire.) Some aid bureaucrats also reason that since the French Congo Republic has oil—a quarter of a million barrels per day, making up 90 percent of exports—and just three million people, it shouldn't need much outside help. I decided to sink a few thousand dollars of my savings into this speculative journalistic venture, and took the Air France jet to the Congolese capital Brazzaville.

Eva Joly's story already illustrates how oil can generate great balloons of mischief money, spreading corruption around the globe. In Brazzaville I was to see another very dark side of this, and of the Elf system.

Oil officials routinely say that they want political stability, and a peaceful, efficient, low-cost operating environment. They *do* mostly want these good things—certainly on a personal level. But do oil companies really *need* these happy conditions? Academic research has found that countries heavily dependent on primary commodities like oil have a greater risk of civil war. One famous study found that states dependent on oil and minerals face a 23 percent chance of civil war in any five-year period, while those with no natural resource exports face only a 0.5 percent chance.[1]

Once again, extracting oil is different from "normal" businesses like tourism or making electric razors. Nobody would destabilize or attack a country to capture its tourist revenues; this would be like using hand grenades to catch the goose that lays golden eggs. Oilfields are not like golden-egg-laying geese but are instead are like solid gold eggs themselves, buried underground; it may be feasible to blast them out with explosives—even if you damage them you will still get most of what you want. With oil, your profits depend less on political stability or a healthy business environment, and depend more on geology, extraction technology, and world oil prices. West African oil has pumped steadily through decolonization, nationalizations, Marxism-Leninism, coups, and civil wars since it was first discovered in the 1950s, and new leaders tend not to abrogate contracts: they know that if they do, their assets risk being caught up in litigation in foreign courts. In fact, oil contracts are safer when African governments feel rather weak and need friends; but when governments grow stronger (when oil prices rise, civil wars end, or new partners like China appear), African governments develop confidence to challenge the companies. A recent spate of resource nationalism around the world, from Venezuela to Chad to Iran, is the result of higher oil prices, and new friendships in Asia.

Companies mostly resist the consequent temptation to weaken governments, for they might be punished if they get caught. But Congo provides a shocking example of where oil interests actively helped tip a country into civil war. It also provides a rare case where an American oil firm, following a visit by U.S. Vice President Al Gore, tried to muscle in on France's jealously guarded territories in Africa, with unintended but terrible results.

Brazzaville is a moldering tropical sweathouse bursting with flowers, birdsong, and thick vegetation, all flourishing vigorously in this hot, soupy tropical air. The center boasts elegant French cafés and wide, pretty tree-lined boulevards filled with smoky Russian vehicles, ramshackle taxis, and new BMW coupes. Bullet holes from the oil wars have mostly been fixed up, but you can still see the damage. From the Les Rapides bar on the Congo River's right bank you can see Kinshasa, the DRC's capital, in plain view opposite, but the river here is so wide that on hazy evenings its yellow lights are only dimly visible. Canoes and larger motorized craft dart between the two cities, dodging matted reeds and lilies that march steadily down from the giant forest basin upstream. Boating is no joke: lose engine power and you will drift down toward vast, brown, boiling cataracts. Kinshasa and Brazzaville are siblings but rivals, joined by family and ethnic links, and by common languages like French and Lingala—a mix of old African tongues, French, Portuguese, and even English ("miliki" is milk, for example), which river traders adopted in the nineteenth century and which spills across the borders. Thrilling dance music from both cities is wildly popular across Africa. Kinshasa was where Muhammad Ali beat George Foreman in the famous "Rumble in the Jungle" in 1974.

In 1482, the Portuguese explorer Diogo Cão wrote of encountering a river so big that fresh water fanned out for miles around its mouth. For millions of years the river's organic sediments settled into this huge fan, which stretches from north of today's coastal oil town at Pointe Noire down to Angola, and the sediments became crude oil. Colonial mapmakers gave Congo-Brazzaville the longest coastline, so it got most of the oil.

To understand the effects of Congo's oil, I wanted to meet André Milongo, who was prime minister from 1991 to 1992. I tracked him down on my second trip to Congo, in 2003.

Milongo's house lies over a short bridge that cuts his suburb off from Brazzaville proper. At the entrance security men checked my bags, and a guard escorted me through the high outside walls into a broad, tiered garden with flowing Bougainvillea bushes, neat lawns sprinkled with purple jacaranda blossom, and plants with thick, dangling red flowers. Workmen were everywhere, refurbishing his grand white pillared residence. Milongo rose to greet me just inside the entrance. Though nearly 70, he was handsome and sprightly, and wore a collarless brown suit and a thick, loose, chain bracelet. After the pleasantries, he began his story, a Shakespearean tragedy with many human actors, and one huge, shadowy non-human lurking offstage, influencing everyone.

"Oil? Yes," he said. "It is behind all these problems."

His tale began with the fall of the Berlin Wall in 1989. "Marxism was a pretext—a religion," he said. "They ruled Congo with it, but not always in the people's interests. It failed in the Soviet Union, and failed here, too." By 1990 political pressure was rising from the streets like heat haze, and president Dénis Sassou Nguesso, a Freemason who had been in power since 1979, was forced to accept big changes. He remained president but lost most of his powers to Milongo, who was popular partly because he had been out of Congo for ten years, and was consequently not so tainted with the incestuous world of *Françafrique*. Milongo, who had been Congo's first paymaster-general after independence and who studied at France's elite École Nationale d'Administration, had more recently been working in—*quelle horreur*—Washington, D.C., as an executive director at the World Bank. To add to Elf's worries, soon after Milongo became prime minister he was received by President George H. W. Bush at the White House.

Congo was then pumping $2 million of oil each day, but civil servants had not been paid for months, and the World Bank was saying that the Congolese oil contracts were terrible for Congo.[2] People in the streets wanted them renegotiated. So Milongo promised to open up the oil contracts to outside scrutiny. "We were not getting the benefits from our oil," he said. "So I asked for an audit." He chose the American firm Arthur Andersen.

Late in 1991—he did not remember exactly when—some people came to see him. "Men from Elf came to my office and asked me, 'Do you need something?' I think they did not want to spell it out precisely. I thought, maybe they wanted me to sign a text to give them the Nkossa oilfield. It is a big field. A Congolese was with them. He said, 'What can we do for you?' I said, 'No—nothing.' I was not used to this. I could see they were angry. Not really aggressive; more surprised—here was someone who did not accept their proposals. They went away. Only afterward, with these Elf trials, did I see what they wanted. I think they wanted me to act like the previous leaders. They were paying African leaders in cents per barrel.[3] I think they wanted to propose something like this."

An Arthur Andersen official came to see him. "He came and said he had to stop the audit: he and his family were being threatened. He was trembling, right in front of me. He had never been threatened like that in his life."

Milongo did not know it then, but this was just the beginning of the trouble.

For centuries slavers ravaged this coastline and sailed up the Congo river, and from the seventeenth to the nineteenth centuries one and a half million slaves were sent from here in a triangular trade that took slaves to the Americas, American cotton and sugar to Europe, and European goods to Africa. Slavers gave guns and money to local potentates, who ruled from colonial trading posts and drained the tribes of the interior, subverting local politics in ways that are eerily reminiscent of today's oil trade.

Other curious similarities exist. You can find one at Pointe Indienne, a quiet area of coastal bush and farmland, where bamboo thickets tumble down to a pretty beach and where onshore oil wells flare gas that lights up the bush and warm surf at night. An old curator at the slave museum here will show you grisly slavers' fetishes and yellowing diagrams of white doctors probing slaves' mouths and genitals. This was the region's main slave export point, and it was also at this very spot, in 1957, that oil was first found in Congo. Museum documents say they crammed three hundred to five hundred slaves per boat, making them dance to tone their muscles and to stop them slipping into "melancholy," and a good male *nègre à talent* was worth the annual wage of a ship's captain (females fetched 25 percent less). You can play mischievously with this data. Take a tanker captain's wage today of, say, $100,000, multiply by 500, and this values a boatful of slaves at $50 million in today's money—about the same as a million-barrel oil cargo.

After the slave trade, western powers sought to tie down territories. The British explorer David Livingstone described Britain's supposedly noble intentions (guided by the "three C's": commerce, Christianity, and civilization) a bit like how some Americans today believe they should spread freedom and democracy overseas. The French had something similar, described in Thomas Pakenham's classic *The Scramble for Africa:*

> Overseas empire would soothe the *amour-propre* of the French army, humiliated by its collapse in the Franco-Prussian war.[4] . . . A whiff of colonial fever, a panicky fear the door was closing (what the Germans called *Torschlusspanik*), had infected the French public.[5] . . . To redeem France's humiliations in Europe by acquiring a great overseas empire, to develop new overseas markets for France, were the aims common to all French colonialists.[6] . . . [Central Africa] was the chance to find the French equivalent to Livingstone's "3 C's": French Catholicism, French civilization, and French commerce—not free trade, of course, but trade only with France.[7]

A French explorer, Pierre Savorgnan de Brazza, reached what is now Brazzaville in 1880, just ahead of his brutal Anglo-American rival Henry Morton Stanley. This is the top of the cataracts, and the start of five hundred miles of navigable river upstream: a gateway to the heart of Africa. From Brazzaville on the river's north bank, France built a five hundred kilometer railway to the coast, at a cost of an estimated 17,000 African lives.[8] The Belgian king Leopold turned his territory on the south bank into a private fiefdom, a brutally profitable rubber-tapping zone that he called his "magnificent African cake," where up to ten million Africans died.[9] Many of them who fled across the river from Leopold's territories reportedly crossed back later to escape the French rubber zones, where there was, as in Leopold's Congo, an estimated 50 percent population loss. "In their lust for quick profits," one academic wrote, "civilized men turned savage."[10]

In France's colonial archives[11] I found a yellowing letter from 1888, written by a French station chief who was indignant about theft from a French trading post:

> It was decided to use energetic methods to chastise the guilty, who come from the same villages. M. Letellier. . . . only supported by three or four men, taught the guilty village a severe lesson, without being injured and without having any man wounded. The village of Benza-Bembé was razed, and 20 pirates who lived there were killed. We have shown our neighbors that we know how to administer justice.

As colonial administration solidified in the twentieth century, the abuses abated as French administrators rolled back the private company agents' arbitrary justice. Later, when the Nazis occupied Europe, Charles de Gaulle declared Brazzaville the capital of Free France, and after the war it became the administrative center of French Equatorial Africa, a huge territory stretching up to the Sahara desert, and including today's Congo, Gabon, Central African Republic, and Chad.

The French, like the British, divided Congo's people to rule them: they drew the army heavily from the north, while Milongo's Kongo-Lari people, and others, were encouraged to be businessmen and bureaucrats. The numerically weak northerners allied with France, which helped them to stay in power for much of the time after independence in 1960. Congo adopted Marxism-Leninism and flew in Cubans, East Germans, and Russians, and broke off relations with America in 1965. Elf was even temporarily ejected, but after a successful coup in 1968 they were eventually invited back, and Elf-Congo was created in 1970, the year Congo's official Marxist-Leninist experiment began.[12] It is a remarkable testament to the power of Foccart's networks that this officially Marxist-Leninist nation did not, or could not, pull itself far from the francophone orbit.

Congo was steadily sewn into the strange Franco-Gabonese meld (and Omar Bongo later married President Sassou-Nguesso's daughter). If

Gabon was an outgrowth of the French state, Congo was a painful appendage attached to it: a second chamber in the French politicians' oily offshore slush fund. A Congolese academic tried to explain the relationship by drawing me a fried egg, with Gabon the yolk, Congo the white, and France as the frying pan. Brazzaville and Libreville each have giant Elf towers, similar-looking oil ministries, and offices of the shadowy former Fiba bank, the pillar of the Elf offshore edifice that snaked Congo's, Gabon's, and Elf's money around the world's tax havens.[13]

In a book, two French journalists explain Bongo's role: "He defends and promotes Elf not just in Gabon, but in Congo. He has uncontested leadership over his francophone neighbors. Is he a friend of France and its leaders? Much more: he is French, from his toes to his hair."[14]

When Milongo came onto the scene, nasty ethnic genies began to fill the new political spaces that emerged after the Cold War. Politics was a messy Nigerian-style three-way fight, between Milongo's ethnic Kongo-Lari brethren, another southern group sometimes known as Niboleks, and Sassou's Elf-friendly northerners. These groups scrambled for power, fostering instability and reckless corruption. Politicians began to feel exposed in the widening political spaces, and began to build ethnic militias. Violence began to break the tension.

At one point, in January 1992, soldiers in trucks started heading toward the bridge from Brazzaville to Milongo's house. Someone raised the alarm, and his supporters went to the bridge. "What saved me was that I lived here," Milongo said. The soldiers tried to get through, and a fight broke out. But the soldiers backed off.

"It was a coup, to get rid of me. It was the army, and behind them, Elf," Milongo said. "Elf! A senior Elf man flew in, on an Elf plane, to organize it. I never thought they could organize a coup d'état for those reasons. They talked all night, then agreed on a coup. They would get rid of me, to stop the political process. They did not want the changes to continue because Congo would have discovered the secrets." (Years later, a former Congolese finance minister in the Elf trials in Paris alleged a plot by Elf in January 1992 to support an attempt to topple Milongo.)[15]

Before the coup Milongo had received André Tarallo, the head of Elf-Gabon, for dinner in his house. "He was great. He didn't make any approach to me," Milongo said. (In the Elf trials, Tarallo denied backing a plot and said that Elf's official policy was always to support the leader in office. A company spokesman subsequently said it was "beyond our imagination" to suggest that Elf would back a coup rather than deal with a government in power.)

Yet the French interests had another, more subtle, control mechanism: debt, which had blossomed in the 1970s oil boom, just as it had in Nigeria and other oil states. Congo launched a five-year plan in 1979, based on huge borrowing, to "break the dependence" on France and "reconquer the nation," and that year it took out its first oil-backed loan, via Elf.[16] By 1980, Congo's debt was larger than its national output,[17] and Congo's three million people, with among Africa's highest average incomes, had the highest debt per capita on the continent.[18]

In the oil boom Congo had expanded an already capacious bureaucracy inherited from French colonialists and Soviet-styled central planners, but when oil prices fell in the mid 1980s, Congo could not print money to cover the bloated payroll, since its CFA currency, shared with Gabon, was controlled by France. New loans were needed to pay off the old ones, and an addiction grew.[19]

Falling into arrears on salaries to the civil service was widely taken as a potent sign of political weakness, which in turn was a boon for Elf and other creditors: Congo's leaders, an Elf official on trial in Paris later explained, had to "put up with" very high interest rates on Elf loans because they were so weak.[20] Elf rescheduled Congo's debts in exchange for new oil concessions,[21] at bargain-basement prices, and provided new loans at exorbitant rates. Bribes flowed through Bongo's Fiba bank and the tax havens, while politicians looted the treasury. Tax havens are like one-way filters for money, sucking African economies dry: once rulers get their cash out, they know it is safe.

Under IMF-inspired privatization, French interests converted Congo's debts into discounted shares of state firms. French businesses and networks gained footholds, building import, forestry, and gambling empires. French official credits,[22] along with painful layoffs and wage freezes, produced fresh money that ensured that the private French creditors, including Elf, got paid back. The spiral of debt and salary arrears deepened, further weakening the politicians. "Losses for the state were turned into private gains for the company, Elf officials, and the ruling elite," wrote the transparency campaigners Global Witness. "The company created conditions of deliberate indebtedness through oil-backed lending, progressively securing its hold on the country's internal politics."[23]

Milongo had inherited seven months' salary arrears in 1991, plus unpayable debts of nearly $5 billion. On a visit to Paris, just after becoming prime minister, he asked for help. This had routinely been given in the past, but this time the French snubbed Milongo, who published a direct appeal to Mitterrand in January 1992 in a French newspaper, pleading for French support.[24] Instead, he had to borrow over $200 million from Elf and its Italian partner Agip, on ruinous terms. The promise to "reconquer the nation" from France had become a joke. "Congo," le Floch-Prigent said in court, "is under Elf's control."[25]

Yet Milongo was only a transitional caretaker, whose job was to prepare for elections in 1992. These polls presented Elf with another problem: their preferred candidate, Sassou-Nguesso, was deeply unpopular. So Bongo stepped in with an alternative: Pascal Lissouba, who was a former prime minister, a Freemason, and a Congolese ethnic cousin of his. Elf backed Lissouba's campaign, and he won.[26]

Yet Lissouba, once in office, faced the old problem. Congo's oil money was disappearing into the black hole of old oil-backed loans, with little left over. "The treasury was empty," said Patrice Yengo, a Congolese academic. "Lissouba found the state did not properly exist."[27] Not only were salaries four months in arrears, but part of the army was still loyal to Sassou-Nguesso, who had mysterious sources of money. So Lissouba began to arm his own militias, and he reneged on a pre-electoral pact to appoint Sassou's allies as ministers of oil and finance, the most dangerous posts in government. Perhaps even worse for Elf, he invited Amoco, Chevron, BP, and Conoco in for talks.[28]

The lid on the Congolese pressure cooker began to rattle. Barricades went up in the streets, there was sporadic shooting, and even a failed coup attempt. Lissouba asked Mitterrand, Elf, and even Bongo for money, but they all refused. Finally, he attacked Elf's contracts, raising its royalty rates from 17 percent to 33 percent. "Lissouba went to look for money," Milongo remembers. "He asked for advances. Elf refused. They supported Sassou and they wanted to punish Lissouba. Lissouba had gone too far."

By now civil service salaries were six months in arrears, and rising, and Lissouba's government was being scornfully dubbed *Gouvernement ya Nzala* in the Lingala language: a Government of Famine.

Lissouba finally went for the nuclear option: an American company. Occidental Petroleum (or Oxy, as it is sometimes known) was founded by the late Armand Hammer, a mysterious deal maker who counted Stalin, and the family of then U.S. Vice President Al Gore, as friends. In April 1993, Oxy agreed to give Lissouba $150 million, routed outside the normal Franc zone banking channels so that France could not block the cash.[29]

Oxy, too, knew that Lissouba was desperate, and it extracted staggering terms: 50 million barrels of future oil, from fields operated by Elf and the Italian firm Agip, at just *three* dollars per barrel.[30]

Elf was enraged. "Elf's arrangements were impossible," the Paris-based *La lettre du continent* wrote, "with the Oxy fox in the chicken coop."[31] The head of Elf personally called up Occidental officials, and insulted them.[32]

A Congolese official remembers the moment when the loan came in, immediately triggering the civil service salary payments—just a day ahead of legislative elections. "The president said, 'You will be paid tomorrow!' We forgot the election and went to the bank, to get our money. *Gouvernement ya Nzala*—this insult fell to the ground. Lissouba had saved himself." The loan was decisive, and Lissouba's party narrowly won.

The next day, Brazzaville had crystallized into fiefdoms held by militias—Cobras, Ninjas, Cocoyes, Mambas, Aubevillois, Zulus, and more—and tanks were rolling into the streets. In the ensuing fighting, perhaps 2,000 people died, and 300,000 civilians fled, as drug-addled thugs raped, decapitated, and skewered their way through Brazzaville's suburbs. Lissouba clung to power, but Congo's democracy was shaken to its roots, as guns flooded in and were dispersed among Brazzaville's households. New fighting erupted in October, but once again Lissouba survived.

Congo calmed down again, yet the economic crisis persisted. Elf bought out Congo's 25 percent stake in Elf-Congo for $40 million, a fraction of what it was worth. By now, following the Rwanda genocide and the start of Eva Joly's investigations, French Africa policy was in chaos, as modernizers battled with the old guard. By 1995, salaries were 13 months in arrears, and debt service—much of it owed to Elf—was taking up nearly two-thirds of the budget.[33] Nobody knew where the oil money was going. "We were in government—we knew nothing about it," one of Lissouba's senior ministers later told me. "The *dossier pétrole* was the preserve of three or four individuals, only."[34]

Ordinary people turned to informal markets to make ends meet, but the politicians tried to control these, too. "A shadow state develops," one academic account explained of this time, "as the administrative capacity of the state collapses . . . a vicious circle emerges: as the official system collapses, people turn to unofficial activities, and society and economy become more violent and chaotic."[35]

In December 1995, American Vice President Al Gore stepped off a plane and met Lissouba for talks, emerging to tell journalists that "under the wise and courageous President Lissouba, Congo is becoming a hope for liberty and a symbol of democratisation."[36] The same day, Chevron was awarded a new oil license. Six months later, Occidental sold its oil interests in Congo back to the government for $215 million, making a tidy profit.[37]

Another coup plot was uncovered, there were army mutinies, and a rampage by Sassou's Cobras. Lissouba began to get intelligence about arms flows from Gabon to Sassou's home in northern Congo,[38] and this spilled out into *radio trottoir* (literally, "pavement radio," or the rumor mill), adding to the fear. Lissouba banned public protests, and brought in Israelis to train his militia. Congo was heading for another explosion.

Early in the morning of June 5, 1997, eight armored cars surrounded Sassou's compound.[39] The defenders fired a rocket, which hit the nearest vehicle; then a second rocket took the tracks off another vehicle idling by a gas pump, and fighters opened up with AK–47s. Brazzaville's sleepy residents, waking up, knew this was not just skirmishing. This was the war they had been fearing.

Milongo, who was the speaker of parliament at the time, was at home that morning. There was no fighting in his suburb, but he admitted being afraid. "Arms were coming in," he said. "We think it was Elf. Sure, there were ambitions: Lissouba and Sassou. But you look right to the bottom of it and you will see that this was all about control over the oil."

Once more Brazzaville divided into three: Sassou's northerners, in northeastern Brazzaville, fought President Lissouba's southerners in the north and center of town, while a third warlord, from Milongo's ethnic suburbs, simmered on the sidelines.

Militias again sacked Brazzaville, each foraying into enemy territory to pillage, then ferrying booty back in armored cars. By July the death toll had reached 2,000, and the war began to bog down.[40] Lissouba's men flung bombs from Ukrainian helicopters over Cobra suburbs, killing many but achieving little of military value. News of atrocities trickled past half-interested foreign news editors: 10 dead in Makélékélé suburb, 15 rape victims in the hospital, faceless families fleeing who knows where in the confusingly named Congo Republic. Amid the carnage, Fiba bank was buying back Congo's debts on secondary markets, at colossal discounts to their face value.[41]

Lissouba remembers getting a call from Chirac, who urged him to appoint Sassou vice president and head of the army. "I said that in our constitution this is not foreseen. We have a prime minister, we have no vice president, the army is under the president, under me. It is our constitution. Chirac told me, 'Your constitution? Put it in the rubbish bin.'"[42]

Sassou got help from Rwandan *genocidaires* and Mobutu's former presidential guard. Lissouba unwisely linked up with Angolan UNITA rebels, infuriating the regional superpower. Then, in mid-October, the dam broke. Sassou's Cobras smashed into Lissouba's suburbs, taking the airport and the presidential palace, while Angola thrust an armored column, backed by fighter-bombers, north into Congo.[43] Angola's disciplined, battle-hardened, and better-resourced soldiers easily overran Lissouba's rabble, and Sassou was soon back in power. His Cobras treated themselves to a bloody celebratory rampage.

One of Lissouba's ministers[44] was in Pointe Noire when the Angolans came. "A minister and his wife had been killed and put in a freezer. I left my hotel: 40 years a civil servant, and I had no time"—he picked up a spoon—"to take this much." Disguised as a doctor, he headed for the port, which was filled with wounded, and got onto a boat. But someone had recognized him. "They knew someone important was on board. The boatmen—they were Yorubas from Nigeria—hid me in the pilot's cabin; then the soldiers came: a *contrôle* [search]. The boatmen took me to the motor room and gave me overalls. Another *contrôle:* Angolans, Cobras. All the passengers off! They put me in a wooden box, to hide me. I stayed in the box for five days, and I did all my bodily functions there. The soldiers came back, went away, came back. Eventually they decided the information was false and we sailed for Benin. At three A.M., they told me I was safe."

Relief workers found Brazzaville in ruins. Ten thousand reportedly died,[45] but the true toll is unknown. An aid worker who saw the Rwandan genocide told me that while the killing in Rwanda was far worse, Brazzaville was in worse shape, because the Rwandans were poor and used bullets sparingly—in oil-rich Congo, the militias had bullets by the truckload,[46] which they sprayed around with abandon.

The day after Sassou was sworn in as president, he received the head of Elf, and a month later he dined in Paris with Chirac, who declared himself delighted with the outcome. All through the fighting, the oil had flowed uninterrupted.

Circles in Paris began putting out misinformation, that it was the CIA that had backed Sassou.[47] In fact, the guns came from French right-wing circles via Gabon; Most shockingly, it seems, from testimony in the Elf indictment and elsewhere, the company supported both sides in the fighting,[48] including $52 million paid through Fiba bank[49] for helicopter gunships for Lissouba, to be paid back in Congolese oil.[50] "Arms were sold to Lissouba, thanks to Fiba," said Le Floch-Prigent, the former head of Elf. "Thousands of Congolese died, and now the survivors must pay for the arms that killed their loved ones."[51]

Congo was soon in convulsions again, with scattered fighting in 1998 and 1999, in which maybe 60,000 people died of wounds,[52] hunger, and sickness; ten times that number[53] fled their homes. Foreign news editors barely noticed. Since then the fighting has ebbed, with occasional outbursts. Sassou has consolidated his position and, especially after the rise in oil prices from 2003, has become more confident, standing up to Elf's successor Total. Milongo is still active in politics; he was to be Sassou's main challenger in a presidential poll in 2002 but he withdrew, alleging fraud. He still criticizes Sassou, but he is no firebrand.

Now, as France's grip slackens, the old Franco-African networks are mutating into more privatized, fragmented forms. "Today, more and more, [the networks have become] lobbies," said a tellingly titled book, *These Africa Gentlemen: From Networks to Lobbies*,[54] "groups who still, obviously, count on the politicians and civil servants, but who only now wave the flag in pursuit of their own financial interests."[55]

The debt mountain has been mutating, too. By 2005, Congo had over $9 billion in foreign debt, unpayable at well over $2,000 per inhabitant.[56] The addiction remains in place: Congo is still taking out new emergency loans from French and British banks,[57] sometimes repaying at five percent *a month*—or more.[58] Congo has promised the IMF to stop doing this, then it has broken the promises.

French interests have sold Congolese debt to American offshore funds such as the British Virgin Islands (registered U.S. fund Walker International), the Cayman Islands (registered and New York-based Elliott Associates), and FG Hemisphere, also of New York. (This is a perfectly legal business practice.) They have been pursuing Congo and a French bank (BNP Paribas) in U.S., French, and British courts. Secret grappling is taking place in the same arcane world of lobbyists, offshore lawyers, and tax havens that foxed Eva Joly. Elliott Associates employed Kenneth Adelman, a former assistant to U.S. Defense Secretary Donald Rumsfeld and a member of the Pentagon's defense policy board, to fling mud at Sassou ahead of his meeting with President George W. Bush at the White House in June 2006. Congo, for its part, has employed the services of Michael Ledeen, a former advisor to Ronald Reagan's Secretary of State Alexander Haig and a leading right-wing scholar and high-profile backer of the Iraq war.[59] The hedge funds allege that Congo's leaders cooked up private conspiracies to disguise oil profits and steal the money.[60] Sassou reacts with fury by wildly calling his detractors, the American funds, " 'snakes in the ocean' and 'vultures,'" and even alleges what has to be an utterly improbable "association of malefactors" involving these funds and the anticorruption watchdog Global Witness, which has also excoriated Congo's rulers.[61]

Congo's state oil company has been putting hundreds of millions of dollars in weird special purpose vehicles (SPVs) in tax havens like the Cayman Islands.[62] The SPVs are designed to be "autonomous in relation to [the state oil company] and the Republic of Congo"; in other words, Congo's oil receipts technically don't belong to it, but to someone or something else, so that creditors can't seize them. Who owns the oil money, then? The only thing that is clear is that these financial gymnastics further separate Congo's war-weary people from their rulers and their oil.

"The Congolese people do not know much about how much our country receives from this black gold, and even less about how the revenues are managed," Congolese church leaders wrote in an open letter.[63] "What it does know is that the price of oil is measured not in barrels or dollars, but in suffering, misery, successive wars, blood, displacement of people, exile, unemployment."

Meanwhile, it is not just inefficient public companies that are being privatized—but the very *functions* of the state. State shipping taxes on oil and timber are paid to private interests, and not recorded in the budget. The national airline collapsed down to little more than a vehicle for collecting air transit fees from foreign airlines; private firms have armlocks on port and shipping facilities, telecommunications, and banks, breaking laws freely or having parliament rubber-stamp new ones in their favor.[64]

As state institutions give way to private interests, Congo's government stands increasingly on just three remaining pillars: first, the internationally recognized sovereignty that legitimizes the oil and banking contracts; second, the state oil company and the oil and finance ministries that manage the financial engineering; and, finally the armed forces that protect the system. Even ongoing low-level conflict is tolerated, as long as it does not threaten the sovereign, extractive core. "Not only do these state institutions survive, but the state begins to hang off these institutions as if nothing else existed," said Ricardo Soares de Oliveira, a U.K.-based specialist in Africa's oil nations. "They deal with the intricacies of oil-backed loans and the oil industries. They become the state. It is not a collapsed state but a privatized state. With a collapsed state, the rulers lose control; with a privatized state, they can even increase the possibility of accumulation."

Today Congo is what the writer Robert Kaplan might have meant when he described parts of Africa reverting to the world of the Victorian atlas, with wealthy trading posts surrounded by what an old map in my living room calls Terres Sauvages—Wild Lands, "blank" and "unexplored." Congo is a Mad Max society, with areas of state control like Brazzaville and the oil town of Pointe Noire surrounded by "useless," lawless territory filled with roaming brigands.[65]

"The people are utterly disenfranchised," said Paul Foreman, the local head of the aid agency Doctors without Borders.[66] "In the city everything is expensive; in the rural areas, there is nothing. Rape and gang-rape are taken to violent extremes; there is a total lack of structure in society, and generalized impunity," he said. "Congo is two nations."

One nation is evident to rich visitors; an attractive world of international airports, satellite dishes, oil rigs, French cafés, and air-conditioned hairdressers, all protected by attack helicopters, decorated with oil money, and plugged into the tax havens.

In Makélékélé hospital in Brazzaville, I find the other world. A cry of pain fills the morning air, from a man hurt in a recent army attack on his village. He is too distressed to talk. In the next ward a 32-year-old man in dirty jeans and sandals lifts his shirt to show me a bloody splash across his back: the aftermath of a helicopter attack. His wound smells terrible. As soldiers came in with the helicopters, he said, nobody had time to collect the village madman, who was howling in the village square. The soldiers removed the madman's head, and moved on.

A local journalist does not see two worlds, but several. On a beer mat he draws a cake, shading each layer. "On the bottom: we, the Congolese. Next up, traders: Nigerians, Senegalese, and Malians. You have rich boutiques: the Mauritanians. Then, Lebanese and Indians; then, higher up, French businesses." He puts a final layer of icing on top, and shades it black. "The Mafia. Congolese, French, all nationalities. They rule!" He adds little arrows, pointing at each layer. "The Chinese . . . they are now coming in at all levels. In photocopy shops, construction, and soon oil. They start on one level; in six months they are up at the next. They will be at the top now for sure." A friend of his expands the point. "The democracies have mafialike things they want to do, but cannot do at home. So they come and do them here, *chez nous*."

The journalist, who does not want me to use his name, takes me to Brazzaville's main rail station, where dark-suited armed gendarmes in front of the station sport new AK–47 automatic rifles. Soldiers in dirty bottle-green uniforms, some with Cobras tattooed on their arms, are lounging against armored rail cars. Men on hand-cranked vehicles shunt back and forth, and next to wagons bearing huge logs trucked down from Congo's northern forests to pick up the armed rail convoys that will take them to the coast for export to China and Europe. The train must past through the forests of Milongo's unstable Pool region, where long-haired drugged Ninja fighters routinely shake down aid workers and any Congolese who venture into this lawless place. Sometimes the Ninjas come to the market nearby: an intricate bedlam of sharp-elbowed hawkers and stallholders, soukous music, diesel engines, and sizzling fish. The market is foggy with flies and wafts with tangy perfumes of tropical cooking and meat from grilled apes, antelopes, and even whole crocodiles, trussed in twine, which people eat with manioc. The Ninjas are loyal to a messianic, armed pastor who wears purple robes and dreadlocks. The police fear them, and rarely disturb them.

Occasionally, they attack the rail convoy; then the line closes and profiteers in Brazzaville sell hoarded fuel at five times the going rate. The army retaliates: villages near the line, blasted by machine gun and cannon fire, become *villages fantômes*—ghost towns—with unprotected mud walls, looted of their roofs, dissolving in the rain. An aid worker remembered one train returning laden with corrugated iron roofs harvested from the Pool.[67] The rumor has long been that hard-liners in the government are scheming with this pastor to keep the military problems alive, for their own twisted reasons.

By the tracks, my guide warns me. "We are being watched, and this evening I will be called, to ask why I brought you here," he said. "This is a *zone de Mafia*."

Later, in my hotel's garden in Brazzaville, a senior accountant meets me. He is well dressed, perhaps very wealthy, but his teeth are rotten; his hands shake, and he stares sourly from eyes yellow with old malaria. "Competition is almost impossible here," he says. "It is a fundamentally criminal structure. The Corsicans, the Mafias: they control it. Customs is in private hands, though Article 11 of the constitution says it should not be." The Dutch disease has made local industries uncompetitive, leaving only firms making low-value heavy or bulky items too expensive to import. "There is beer, sugar, timber, and cement," he said. "There is nothing else."

A strange oil deal had just been finalized with Total, Elf's privatized successor company, under which a medium-size oilfield it owned was transferred to Congo, then shifted immediately to a mysterious private company.[68] "It is another loss for Congo," the accountant lamented, rolling jaundiced eyes. "But—please, you tell me—what can an asphyxiated country do?"

OBIANG NGUEMA

WHAT CARING NEIGHBORS DO

As President Obiang Nguema's limousine approached the Congresses Palace in Bata, soldiers brandished whippy sticks to part the worthies in green caps and white party T-shirts. For months, officials had been organizing this, ordering tie-pins and leather-bound folders, and faxing flowery invitations. This, the ruling party's third ever congress, was a celebration of Equatorial Guinea's full-blooded entry into the oil age.

It was July 2001. The American firm Triton was starting up a big offshore oilfield, just 14 months after discovering it—a world record at such water depths. Mobil was pumping fast from its giant Zafiro field. The government, facing an uncertain future, also wanted to get the oil out fast. "What production profile the government wants . . . is determined by the security of the government," said Max Birley, vice president of Marathon Oil Equatorial Guinea.[1] "An insecure government will want the oil reserves exhausted as quickly as possible." By now, Equatorial Guinea's half a million people already boasted the world's fourth highest production per capita, more than Saudi Arabia or Iran. Though less than a quarter of the value of Equatorial Guinea's oil was flowing from the oil companies to the treasury,[2] an exceedingly low share by regional standards, this still meant a flood of money that was already leading to inflation, big salary rises, and what the IMF called "a breakdown in budgetary discipline."[3]

Kicking off the congress, the planning minister handed me a paper predicting that the economy would grow faster than any other in the world that year: 71 percent, upstaging even the IMF's dazzling 53 percent forecast. Men in suits and supercilious ladies trooped up to the stage to praise president Obiang and his wife Constancia, the Mother of the Nation. "One man! One woman! One party!" speakers roared, to raised fists and cheering. "This will be the Kuwait of Africa," yelled one, and the hall erupted. He was mobbed by party cadres like a football star. Constancia, a commanding speaker, chided all males for their generally bad habits; the men snickered and clapped. Obiang took the podium, softly emphasizing his power. "There is practically no opposition in our land because nearly all political parties, backing the [ruling party], support the same principles." In the audience, behind impassive diplomats from North Korea, sat a Spanish businessman in dark glasses who seemed to say and do nothing for three days. A big-boned American, Bruce McColm, sat nearby. Two women from the U.S. Congressional Black Caucus, sporting white party neck scarves, raised their fists.

Outside the big hall, watched by Moroccan bodyguards, a lady was shaking her hips and praising "Hermano Militante" Obiang into an over-electrified microphone, as groups of colorful praise singers chanted and slapped their thighs. Further out, restrained by a low wall and the soldiers, people in tattered shirts and flip-flops stood in dirty brown groups, watching and listening. The chances are that they cannot afford doctors, one in six of their children die before their fifth birthday, and their drinking water tastes of mud.

They were not just missing the bonanza: their lives were being turned upside down, too. Agriculture's share of the economy had fallen from two-thirds of the national output a decade earlier, to a twentieth.[4] A cocoa company official said his profits were collapsing, and he was ceding offices to the oil firms. "Oil has worsened the differences between our citizens,"[5] an Equatorial Guinean U.N. economist, Fernando Abaga, wrote in a paper about the oil boom. "An opulent majority sails in a sea of misery."

Nevertheless, the mood was very unlike that of the introverted little state I had experienced eight years earlier on my first visit. Shops were fizzing with French champagne, Japanese televisions, and Spanish fashion, and Malabo was choked with traffic jams and gaudy tropical palaces. In the Pizza Place downtown, oil workers in jumpsuits mingled with Lebanese traders and local military top brass, while Cameroonian women in sparkly miniskirts sipped Fanta, watching them all. Light from a giant gas flare on the headland bathed Malabo orange at night.

Obiang was not as brutal as his bloodthirsty uncle Macias, whom he toppled in the 1979 coup. Yet reports by Amnesty International and the U.S. State Department still placed Equatorial Guinea among the world's most unsavory dictatorships. International narcotics organizations said that Equatorial Guinea's diplomats used diplomatic bags and other mechanisms to smuggle drugs internationally. In a 1999 book, *The Criminalization of the State in Africa,* three well-known academics concluded that Equatorial Guinea was one of only three states in Africa that could be classified as criminal states.[6] "It's not even a country; it's a confusion," an outraged diplomat said. "A group of people have got hold of this country and are undertaking acts of piracy."

Just ahead of my 2001 trip a Gabonese friend joked about Equatorial Guinea. "Watch out!" he laughed. "The opposition people there have pointy ears." I knew that Gabon's silky elites despised their rougher Spanish-speaking neighbors, but I did not immediately appreciate the cruelty of his comment.

In Malabo I met a short, timid man who did indeed have stunted ears. He told me that in January 1998 three soldiers had been killed by ethnic Bubis, the indigenous inhabitants of Bioko Island, who believe that the oil that lies near Bioko Island is theirs, and resent the daily humiliations they suffer at the hands of Obiang's Fang ethnic group. They had attacked the soldiers with hunting guns and knives, with the vague intention of throwing off the Fang yoke. "They did not have a plan," my disfigured interlocutor said. "They just wanted the world to know what is happening here."

Retribution was swift. Soldiers and Fang civilians dragged Bubis from buses and their homes and beat them; some were executed. At police stations men had their shoulders dislocated; women were raped and made to "swim" naked in mud and "show what they did with their husbands." One woman was told—regarding a fork that had been thrust into her vagina— "from now on, that *is* your husband."[7] The man I met said soldiers bound his hands and feet behind his back then cut off the tops of his ears with scissors, before smearing him with sardines and dumping him in stinging ants. Others suffered "la huevera," a painful torture involving the testicles, or the "hanging bat," where prisoners' arms and ankles were tied behind their backs; from these trusses they were then suspended, blindfolded, from hooks. Arms break from the strain. Some men needed help urinating afterward because they could not use their hands. About 500 people were arrested; 15 were sentenced to death, and 70 got prison terms of up to 26 years for "terrorism" and "undermining state security." Some sentences were commuted, but a few prisoners died in jail after what Amnesty International said was prolonged torture. The government said the mutilated ears—there were several cases—were birth defects.

The Bubis lie on the outermost ring of concentric circles of power here. At the center sits the president; next is his close family, then his sub-clan, then his Esangui clan from Mongomo (a town south of Bata). The Mongomo clan is part of the larger Fang ethnic group that spills over into neighboring Gabon and Cameroon. Kin relationships are central: some say that when two old Fang meet they can talk for hours to establish their relative places in the ancestral tree, up to ten generations back. Those outside the innermost circles, such as the Fang opposition leader Severo Moto, are mistrusted, and Obiang relies on Moroccans and others for his personal security.

An opposition activist, Juan Nzo, put it simply. "La Família. There is nobody else." Others, in private, call them Los Gordos or Los Intocables—The Fat Ones; The Untouchables. "The institutions of state have a phantasmal existence," said an exiled opposition party.[8] "Everything is manipulated to the will of the dictator."

The Mongomo Clan has kept a tight grip since independence, when Obiang's uncle Macias took over. Since toppling him in 1979, Obiang has directed bloody purges every 18 months or so, alleging coup plots that are often (but not always) fabricated.

Plácido Micó, a bespectacled, Spanish-trained lawyer who lives above a grubby café in Malabo, spoke vehemently. "Lots of us have been in prison and tortured. They beat us until they run out of energy. They are animals. The people live like rats; without international pressure we would be dead." He said he'd tried talking to the oil companies, but without success. He was irritated about "Eric the Eel," a local swimmer who put in a comically slow performance at the Sydney Olympics. "They decorated him for being the worst swimmer in the Olympics. They are making a country of clowns, a society of imbeciles. Instead of recognizing this as a failure of the state to provide swimming facilities, they have concocted a triumph. People are laughing at us."

Quaint edicts periodically come from on high. A Malabo resident recalled a morning before an African leaders' summit meeting, when he was woken by a loud banging on his door. It was a government minister who had been told to get their street swept before the dignitaries came. His street was filling with bleary-eyed neighbors with brooms, sweeping. Once, in 2000, half the ministries were relocated from Malabo to the mainland. "In our country, if the minister is not there, the people in the ministry do not work," a top official told me. "We have to move the ministries, with their ministers, to get the work done [on the mainland]."

Obiang is everywhere: on television or on the radio; he looks down from beneath his craggy eyebrows, framed in large portraits that hang in all the offices, which are talismans against bad luck. The charmed few of Malabo can eat out for free, scrawling worthless IOUs to restaurateurs, then melting into the night in air-conditioned vehicles. State media praises him, sometimes a bit too effusively. President Obiang "is like God in heaven. He has all power over men and things," the radio once said. "He can decide to kill without being called to account and without going to hell because God himself, with whom he is in constant contact, gives him this strength."[9]

Yet Obiang is weak in other ways. A foreign businessman remembers how hard it was to get a proposal for a joint venture with Obiang through the cabinet, or Council of Ministers. "With the council of ministers everyone has a say. It has to be unanimous. If you disagree, there is no decision. It took us three meetings to get our centre created," he said. "The council of ministers is really a set of different political bases, and each brings something to the table. Obiang has to weigh whether to alienate a base or not. And often he cannot. . . . They call him the 'Great dictator'—but there are virtually no consequences for a minister disobeying orders. He says: get XYZ done, and a year later nothing happens." Perhaps, like a playground bully, Obiang is fierce because he is insecure.

Obiang is also a joy to interview. I have met him several times, and each time he seemed unconcerned with political correctness or how his comments might play with a western audience. He is tallish and slender, with a soft handshake and bushy eyebrows giving the impression of subdued, tired mirth. When he furrows his brow and wrinkles his nose, perhaps in response to an indelicate question, he is like a sensitive grandfather, pressing his hands together at the fingertips and cocking his head, concerned to set the record straight. In an interview in 2002, together with a French journalist, I asked him about the Bubi ear-cutting. These stories, he said in fluent French, slowly wagging his finger, were lies made up by his enemies in the Spanish press. "It was a terrorist coup attempt, to kill people in power here and create panic," he said. "Thanks to the security forces we found out the information." Then he backtracked. "But there was no torture." Could we visit Plácido Micó, an opposition leader, who was then in prison? He seemed irritated. Wait for the next national festival, he said.

That is the time for clemency. He began talking about foreigners, using the French word *étrangers,* which can also mean outsiders. Such people, he said, cannot be trusted to look after his personal security. We pointed out that he had Moroccans in his bodyguard. "I was talking about *étrangers* locally, not those from overseas." So who, we asked, are these outsiders, not from overseas? "I mean people who are not from my family."

Another time,[10] I asked him why American oil firms had been so successful. "Spain said Equatorial Guinea has no oil. France said Equatorial Guinea has no oil," he said, suggesting that perhaps he thought their failures represented a conspiracy against him. "Then the Americans came. In under six months they found oil."

In 2002 I watched Obiang campaigning for presidential elections (which he won with 97 percent). Moroccans in jeans with submachine guns lounged in front of the stage while men from the Special Riot Brigade flanked Obiang, who wore a white party T-shirt and neck scarf. The American lobbyist Bruce McColm stood, stony-faced, on the podium behind Obiang, who railed at the political opposition. "I do not believe any of them renounced being Equatorial Guinean citizens," he said. "If they do not like Equatorial Guinea, they can go overseas." I stopped a large African American who said he was a businessman, but when I said I was a journalist, he nodded, and walked away. An amiable gent with bad teeth, who was from the information ministry, denied this was a repressive place, citing television pictures of prisoners at the U.S. base in Guantanamo Bay. "Torture, you see," he cackled, "it is normal everywhere. The alternative? Chaos."

Praise singers ululated and chanted. "Si, si, si, Papa, oui, oui, oui." Obiang began to talk about oil. "In the sacred story, when the Egyptian pharaoh had a dream, he said there will be fat cows representing abundance, then thin cows will eat the fat cows, representing hunger," he said. "Now we have passed the era of the thin cows. We are now in the time of the fat cows—which is our prosperity!"

Two months later an article appeared in the *Los Angeles Times,*[11] saying that ExxonMobil and Amerada Hess had been paying Equatorial Guinea's oil revenues into government accounts that were effectively controlled by Obiang, at Riggs Bank in Washington.

Riggs had been open for two centuries and had served (in earlier incarnations) the families of Abraham Lincoln, Ulysses Grant, and Dwight Eisenhower. It supplied the gold for the purchase of Alaska from Russia in

1867. At one time, nearly all foreign embassies in the United States, and half of those in London, banked with Riggs,[12] which promised its clients "utmost discretion." (Jonathan Bush, President George W. Bush's uncle, was appointed CEO of Riggs' investment arm in 2000, in an unrelated area of the bank.[13]) Riggs had provided mortgages on a luxury home owned by Obiang in Maryland, and one in Virginia for his brother Armengol.

The Senate and a federal grand jury began investigating Riggs for possible money laundering, breaching banking regulations, and violating the Foreign Corrupt Practices Act.

Following a call from a senior Senate staffer, Obiang's lobbyist Bruce McColm wrote to the Riggs account manager for Equatorial Guinea. "He wanted to inform me that . . . everyone knows the allegations about Riggs are nonsense," McColm quoted the staffer as saying. "The government of Equatorial Guinea must respond quickly with full force to such articles or they will be believed eventually. The government should not count on the oil companies to defend them, because at the first sign of trouble they will run for cover. The *Wall Street Journal* was thinking about doing a piece. Also, a *New York Times* reporter has been snooping around."[14]

The U.S. Senate Permanent Subcommittee on Investigations—the body that had probed Omar Bongo's millions in Citibank in 1999—set about investigating Riggs for the Equatorial Guinea accounts. They also probed more than 150 accounts held by Saudi officials, to see if Riggs inadvertently provided banking facilities to the September 11, 2001, hijackers.[15] (The investigators only had a mandate to look at this from a policy perspective—to see if certain systems encourage wrongful behavior, and then find ways to improve the system. They are not detectives, mandated to sniff out crimes and punish the perpetrators.)

The subcommittee held hearings in July 2004 and invited officials from Riggs, oil companies, and the government to testify.[16] By then, the American bank regulator had fined Riggs $25 million over the Equatorial Guinea and Saudi Arabia accounts, the largest ever fine of its kind,[17] and the accounts were closed.[18] The Saudi matters contained "very sensitive issues—security issues," as someone familiar with the case put it,[19] and were moved from the Permanent Subcommittee on Investigations to the full Senate committee.

Equatorial Guinea had opened its first account at Riggs in 1995, and became Riggs's largest customer, with balances of around $750 million.[20] Riggs set up 60 accounts and certificates of deposit for Obiang and his family, and for Equatorial Guinean officials and companies. One of these was a shell company owned by Obiang in the Bahamas,[21] whose account took deposits in suitcases of cash in plastic-wrapped bundles, sometimes weighing up to 60 pounds.[22] Money flowed, with help from British and Spanish banks, from the government account to other companies held by his family members,[23] or to companies whose mysterious owners were not revealed.[24] The accounts at Riggs where oil companies paid their dues to Equatorial Guinea had only three authorized signatories: Obiang, his son Gabriel, and his nephew. To withdraw funds, two signatures were needed, one of which had to be Obiang's.[25] Riggs rarely asked about this money, as money-laundering laws required,[26] even as *Parade* magazine placed Obiang below Saddam Hussein on a list of the world's ten worst dictators.[27] "Where is this money coming from?" a Riggs memo wondered in 2001, before answering itself: "Oil—Black Gold—Texas Tea!"[28] Riggs helped arrange financing for ten Eland armored vehicles,[29] and a Riggs employee even siphoned money into his wife's account.[30]

"It does not take a Ph.D. or a degree in finance or accounting," Senator Norm Coleman noted at the hearings, "for somebody to think: hey, you know something? There is something amiss here."[31]

Riggs appeared to have ignored easily available IMF reports outlining missing oil revenues in 2001, estimated at an eye-popping 10 percent of GDP. I once asked Obiang why the IMF was so critical. "The IMF is a purely political institution," he said. "They asked us to give them details of the oil receipts in foreign banks. I do not agree that we give this to the IMF. It is a state secret."[32] (This comment was published in *The Economist* and elsewhere; later an Equatorial Guinea official told a contact that Obiang said these things "because the journalist was annoying him.")

Yet Riggs officials had clearly been worried for a while: an earlier *Los Angeles Times* article[33] had prompted internal bank e-mails attacking the journalist, Ken Silverstein. "The writer seems to have a grudge against the whole world," he wrote. "Equatorial Guinea has never been a 'pariah state' . . . regarding the President of Equatorial Guinea being corrupt, I take exception to that because I know this person quite well. Sir, I wish in due course you will get to know the President of Equatorial Guinea and witness his simplicity first hand."[34]

Riggs did, in fact, arrange for outsiders to brief its staff about Equatorial Guinea. The man they chose to do this was Bruce McColm, the American I saw standing behind Obiang at the party rally, who had established a telecommunications joint venture in 2000 with Obiang.[35] After reading his report, senior Riggs officials lunched with Obiang, then wrote him a letter offering "the best financial expertise. . . . [T]ogether, we can reinforce your reputation for prudent leadership and administration as you lead Equatorial Guinea into a successful future." Senator Carl Levin read this letter out at the hearings, and asked, "How do you write this stuff to a man as abominable as this guy, and known to be abominable? How do you write—how do you, basically, live with yourself?"

The initial Senate investigation focused just on Riggs and its role in money laundering. But as they delved, they ran into another matter. The oil companies.

The American companies had paid money not only into the official oil account, but also to prominent Equatorial Guineans. Amerada Hess paid nearly half a million dollars to a 14-year-old relative of Obiang's to rent office space (though the company said that it inherited the lease when it bought its oil interests from another American oil company in 2001).[36] ExxonMobil leased property from a company controlled by Obiang's family, paying rent into an account controlled by Obiang's wife Constancia. The same company, Abayak, also paid $2,300 for a 15 percent stake in a very small local fuel distribution business,[37] and it owned a major stake in another company that—in partnership with the U.S. firms Marathon Oil and Noble Energy—held 20 percent in an operation to produce liquefied petroleum gas, plus 10 percent of a methanol plant,[38] the world's second largest of its kind. In addition, the Senate hearings found that the American firms ChevronTexaco, Devon Energy, and Vanco all paid for local students, many of whom were the children of local officials, to study in the United States.[39] (It is not clear whether the companies were aware of the students' status, and company officials said their oil contracts required them to make these payments.) ExxonMobil also paid for security services provided by Obiang's brother Armengol. These practices do not infringe the U.S. Foreign Corrupt Practices Act; it seems that American law makes it possible to enter into these kinds of partnerships with foreign officials without it constituting bribery.[40]

The oil companies further argued, with some justification, that so much is owned by the ruling family in Equatorial Guinea that they had no choice but to do business with them.[41] Second, some of them said that they had inherited the ventures from other companies, so were not to blame for them and did not have details about how they were set up.[42] This seems odd, given that oil companies routinely check out potential partners before doing business with them, especially when hundreds of millions of dollars are involved. A senator singled out ExxonMobil for being particularly stingy with facts.[43] I question how hard it would have been for ExxonMobil to find records of these transactions—not least because the company it inherited the contracts from was Mobil. An ExxonMobil official testified in response that it took the subcommittee's work "extremely seriously" and said that it had provided thorough and detailed submissions. "We maintain the highest ethical standards, comply with all applicable laws and regulations, and respect local and national cultures." Officials from other companies said similar things about their companies at the hearings.

John Bennett, the American ambassador to Equatorial Guinea, whom I had met in 1993 just as the oil relationship was beginning, attended the hearings.

"It was about as sad a commentary as one could imagine on U.S. business," Bennett wrote in an e-mail. "All three [Riggs officials] obfuscated like hell, in my opinion . . . the U.S. government's bank regulators [who were regulating Riggs] represented a collective disgrace to a professional civil service. ExxonMobil has not been able to file answers to written questions posed by the committee."

Bennett found it hard to believe that the three companies did not know exactly what they were entering into. "This would have been a fun session—outrageous low comedy—if it were not for the nagging thoughts of [what is going on in Equatorial Guinea]."

In March 2005 the Department of Justice, after its own probe, fined Riggs a further $16 million for failing to prevent money laundering, making the bank the third ever financial institution to be convicted of a criminal violation of the United States' Bank Secrecy Act.[44] Riggs was then purchased by Pittsburgh-based PNC Financial Services Group for about $650 million.

For his part, Obiang responded in characteristically idiosyncratic style. "If there is diversion of funds, I am responsible. That's why I am 100 percent sure of all the revenue, because the one who signs is me," he said in an interview.[45] Later, he added this: "Corruption was unknown in our tradition before . . . bad habits have been introduced by foreigners since oil was found. That is why I am obliged to assume the role of sole national paymaster-general . . . so I can exercise the necessary control."[46] Government officials also said that Equatorial Guinea's money was parked offshore partly because the Treasury building in Malabo was not very secure.[47] Sending the money offshore also probably reflects Obiang's desire to protect "his" money from those around him, keen to get access to the oil money.

Many people believe that offshore skullduggery happens only in places like the Cayman or Channel islands. This is wrong: from other countries' perspectives, the United States, London, and other places with supposedly clean financial sectors are *themselves* offshore centers, using bank secrecy to hide and protect money that flows there. The United States is the biggest tax haven of them all, and a delight for foreign dictators.

American oil companies and their friends have glossed over Obiang's ways. Brian Maxted, Amerada Hess's senior vice president of global exploration, described the government officials his company dealt with as "ethical . . . a breath of fresh air. This might just be the country in Africa that gets it right."[48] Chester Norris, a former American ambassador to Equatorial Guinea (Malabo has a Chester E. Norris Jr. Avenue), said that Obiang "really wants to bring about democracy and improve the human rights record . . . it's already pretty good."[49] Before the Riggs scandal blew up, the U.S.-based Corporate Council on Africa (CCA), an industry association, published a 54-page guide to Equatorial Guinea[50] with pictures of beaming children, water pumps, canoes in the sunset, the American flag, and oil rigs. It says that Obiang's government "has taken significant measures to encourage political diversity and address human and worker rights issues," and has placed a "special focus" on transparency. The CCA listed as its sponsors ExxonMobil, ChevronTexaco, and AfricaGlobal partners—a company with a contract to help Obiang with "investment promotion and political interaction" with the U.S. government.[51]

Other links exist between U.S. politics and the American oil interests in Equatorial Guinea, beyond the well-known friendships that ExxonMobil enjoys in the administration of George W. Bush. Tom Hicks,[52] chairman of Dallas-based Triton, bought the Texas Rangers from Bush and his partners in 1998,[53] and according to the Center for Public Integrity in 2000, Hicks and employees of his companies were Bush's fourth-largest career patron, having given him at least $290,000.[54] Shortly after Triton was bought by Amerada Hess for $3.2 billion, Triton's president Jim Musselman added a boost: "Knock on wood, this country is stable and the president is sincerely trying to improve things. It's not going to turn into suburban Washington, but it could be a model for this part of the world."[55]

The ubiquitous Bruce McColm is a former president of the International Republican Institute and a former director of Freedom House, and, later, a cochair of the Iran Policy Committee.[56] He ran two organizations with the initials IDS: one, the Virginia-based Institute for Democratic

Strategies, a nonprofit group committed to strengthening democratic processes abroad, and International Decision Strategies,[57] a for-profit organization that set up a telecom joint venture with Obiang.[58] The institute, which received support from Mobil,[59] monitored elections in 2000 and reported them as generally free and fair. The ruling party won 95 percent of the seats.

Other observers said those polls were a farce. (To be fair, McColm is a more complex figure than perhaps I've painted; while obviously looking after his own interests, he has probably been a force for good too. Having got into a position where Obiang listens to him, he has been a useful conduit for Western concerns about human rights abuses and a proponent of greater openness.) In an earlier poll in 1996, soon after Mobil struck its huge Zafiro field, I watched soldiers order citizens, lined up and called out by name, to select voting slips from the pile with Obiang's name on them, and put them in the ballot box. Later, Chip Carter, the son of the former U.S. president Jimmy Carter, appeared on television. He said he was in town on a private visit (unconnected to his father's well-regarded Carter Center.) "I came here to observe the elections and to see some old friends," he said. "I think the election was transparent and the people were voting with their convictions. It went very well, according to the will of the people." Obiang won with 99 percent. With another journalist, I went to try and find Chip in a building near the presidency where he was staying. A door opened to the sight of chandeliers and marble, and a dapper manservant with dinner jacket and white gloves. "Mr. Carter," he informed us, "has gone fishing with the president."

These links are all examples of the merging of African oil money and western politics. Yet there is another matter, raising separate questions about America's role.

First, a brief aside. In March 2004 I was researching some oil dealing in neighboring São Tomé by the Canadian company Energem, formerly Diamondworks. I knew Diamondworks from Angola: in 1998 they took me on a Boeing aircraft full of foreign investors and stock market analysts to their mines in the badlands of the northeast to drum up support for their share price, which was then languishing below a dollar. I remember the company's chief financial officer leaning over a seat in front of me and telling me that theirs was "easily a $10 stock." On the plane I briefly met a regal-looking British gentleman, in a starched short-sleeve shirt and sharply ironed Chinos, who had helped set up the security for the Diamondworks mines. He was Simon Mann, an upper-class British mercenary who had helped create Executive Outcomes, the company that had assisted the Angolan army in turning the military tide against UNITA in 1993–1994. Diamondworks[60] had won the diamond concession following this military contract. They flew us by helicopter over impressive diamond mining earthworks on the roiling Chicapa river; we saw blackened villages where a "nest" of miners and their families had recently been forced out in a military operation. Under an Acacia tree an overweight, sunburned Afrikaner with a big beard served us an exquisite prawn lunch, while Mad Max black frogmen with rickety air hoses dived nearby in the river for gems. Later a friend in Luanda who moved in "security" circles mocked the operation and parodied a senior Executive Outcomes operative he knew, ranting in a posh British accent about "fucking jungle bunnies"— referring to Angolans.

My friend had a serious point. The white mercenaries were cocksure of their superiority over Africans, but they were surely mistaken: Diamondworks, by virtue of its links with Executive Outcomes, would be UNITA's top target. A few months later UNITA overran the mine I visited, destroying Diamondworks' balance sheet. The defunct firm was bought out, changed its name to Energem, and said it had shed all links with the mercenaries (some of whom went to work in private security outfits in Iraq). It was this new company, Energem, that had done the oil deals in São Tomé that I was investigating, and which were so controversial that the government there nearly collapsed. I am not suggesting at all that Energem had any involvement in what was to come next—I am only noting a coincidence of timing.

Around that time, I heard from a contact about plans for a mercenary coup in Equatorial Guinea. I found this hard to believe: if *I* had heard about a plot, Obiang would surely know, and would take steps to thwart it. Then another contact at a meeting in London, attended by officials from the U.S. State Department and the British Foreign Office, heard an oil company official, citing a security consultant's report, also talking about possible future action of this kind in Equatorial Guinea and São Tomé.[61]

Some days afterward, Reuters asked me to investigate a wave of arrests in Malabo, involving foreigners. I assumed it was just the latest of Obiang's purges. Before oil came, these were usually aimed at those on the outer rings of power: Bubis, or Fangs who were not Obiang's close relations. But as oil grew in importance, competition for cash intensified at the center, and suspicion fell increasingly on the inner circles; this was being reflected in prison sentences, beatings, flights to exile, and faked "suicide attempts" of even close family members of Obiang.

So, one morning in March 2004 I called Equatorial Guinea's information minister. I had just seen a Reuters news item about the detention of more than 60 supposed mercenaries in Zimbabwe on a U.S.-registered Boeing. I asked the information minister for fresh details about the arrests in Malabo. He gave me some, then added this: "There is more. Some 15 mercenaries have been arrested here in Equatorial Guinea, and it was connected with that plane [seized in Zimbabwe]. They were the advance party of that group." The mercenaries, who included black and white South Africans, a German, and Kazakh and Armenian citizens, were acting, he said, on behalf of a Lebanese businessman named Ely Calil who was the "godfather" of the opposition leader Severo Moto. I had met Moto twice and I knew he hated Obiang. He was once arrested in Angola in 1997 with a boatload of second-rate mercenaries in a ham-fisted coup attempt, and had somehow escaped to form a self-proclaimed government in exile in Madrid.

"Can it wait?" the Reuters West Africa boss asked when I called with my scoop. His hands were full with other stories. I insisted that it could not. The story went out on the Reuters wire, setting off a media frenzy that would grip the political establishments in Britain, South Africa, Spain, and the United States.

This schoolboy farce has been heartily chewed over elsewhere, and I will offer only a short version here. But the most interesting question for me is this: Who was the original mastermind of this plot?

Smokescreens were soon puffing out from the mercenary-infested zoo of military and "intelligence" sources in South Africa, Britain, and the United States. In one version, some Americans had contracted the men to snatch the former Liberian warlord, Charles Taylor, from house arrest in Nigeria for a $2 million reward. In another, they were providing security for a mine in the Democratic Republic of Congo, or they were headed for Burundi, or to some mining operations in Sudan. Or Obiang had fabricated the coup to justify a crackdown. I knew a couple of these officials from my earlier times in Angola, where I believe they had been deliberately spreading misinformation.

The world's media hunted for clues, and lawyers and mercenaries, keen to portray their versions of the story, fed journalists with documents. "It is potentially a very lucrative game," said one that was written before the coup attempt. "We should expect bad behavior; disloyalty; rampant individual greed; irrational behavior (kids-in-toyshop type); back-stabbing, bum-fucking, and similar ungentlemanly activities."[62] They would put Severo Moto in charge, it said, then use local intelligence services to dig up enough dirt to weaken Moto, allowing them to set up what was to be their own private oil colony, modeled on the old British East India Company.

A confession handwritten by Simon Mann, the mercenary I had met in Angola and now in jail in Zimbabwe, outlined how Calil had asked him to research Equatorial Guinea. Here is a condensed part of the transcript, outlining a plot that took shape just as the Riggs probe was getting underway:

> He mentioned EG [Equatorial Guinea] to me and asked what I thought about it. I knew nothing and said so. He asked me to do some background and meet again. After two weeks we met again. I now know that the situation in Equatorial Guinea was very bad. I met Severo Moto in Madrid. He is clearly a good and honest man. He had studied for priesthood. I was also struck by just how bad Obyiang [sic] is. Moto introduced me to General Sargosa who had once been head of Obyiang's security. Sargosa left Equatorial Guinea and joined Severo Moto after Obyiang raped Sargosa's wife in front of him. At this stage they asked me if I could help escort Severo Moto home at a given moment, when simultaneously there would be an uprising of military and civilians against Obyiang. I agreed to help.

True or not, the stories left a strong impression on Mann. He then described how he and his partner Nick du Toit tried to get weapons and set up a business partnership with Obiang's brother Armengol in Equatorial Guinea, as a cover for the operation. They reckoned that 75 men were needed to "escort" Moto to Equatorial Guinea, and they contacted a Colonel Dube and a Martin Bird in Zimbabwe for the weapons, telling them they needed them to protect a mine in Congo. "We met Martin Bird at the Cresta Hotel," Mann wrote. "His wife was also present. I found this most off-putting. We met Colonel Dube, who seemed negative. . . . When I tried to explain the cover story for why we wanted the weapons, he was not interested. When I tried to show him where the mine is on the map, he did not look."

Two weeks later, at another meeting in Zimbabwe, Mann's partner asked to add 2X SA7 missiles to the order. "This seemed unnecessary but, more importantly, dangerous because to ask for such sensitive items might raise the alarm and compromise the whole deal," Mann wrote. "The Spanish prime minister has met Severo Moto three times. He has, I am told, informed Severo Moto that as soon as he is established he will send 3,000 Guardia Civil. I have been respectfully been told that the Spanish government will support the return of Severo Moto immediately and strongly. I can state categorically that I have no links with South African, United States, United Kingdom, or Zimbabwe security intelligence services. I say this because I have been asked the question."

Mann later said that he wrote this following days of torture. "They dictated to me what I should write and at every instance I objected I was subjected to further torture and assaults. . . . I was extremely distressed, disorientated and extremely vulnerable." [63]

Mann also tried to sneak a letter out to his wife, which was intercepted. It mentioned "Scratcher" (alleged to be Mark Thatcher, the son of the former British Prime Minister Margaret Thatcher), a J. H. Archer—allegedly the former British Conservative Party politician Jeffrey Archer[64]—among others. Archer denies involvement,[65] but Thatcher, no friend of South Africa's ruling African National Congress (his mother once called Nelson Mandela a "terrorist"), was arrested at his luxury home in Cape Town. The letter said:

> Our situation is not good and it is very URGENT. [The lawyers] get no reply from Smelly [alleged to be Ely Calil—who strongly denies any involvement in the plot], and Scratcher asked them to ring back after the Grand Prix race was over! This is not going well. I must say once again, what will get us out is MAJOR CLOUT. We need heavy influence of the sort that . . . [a former advisor to Margaret Thatcher and several defense companies] can exert and it needs to be used heavily and now. Once we get into a real trial situation we are fucked. It may be that getting us out comes down to a large splodge of wonga [British slang for money]! Of course investors did not think this would happen. Did I? Do they think they can be part of something like this with only upside potential—no hardship or risk? Anyone and everyone in this is in it—good times or bad. Now it's bad times and everyone has to F-ing well pull their full weight. . . . If there is not enough, then present investors must come up with more.[66]

But they could not wriggle free. Mann and his coconspirators stayed in jail. Equatorial Guinea pursued legal actions against the plotters in South Africa, London, the Channel Islands, and Beirut, largely without success (Mark Thatcher was forced into a plea bargain, involving his admitting that he provided funds for a helicopter, and he was released). International law is unhelpful to those seeking to prove a coup plot; and bank records were lodged in the tax haven of Guernsey, which did not cooperate. "This is not an area that lends itself to litigation," said Antony Goldman, an analyst who knows Equatorial Guinea. "Whatever happened, the lawyers have had a field day."

Who was ultimately behind this?

France's close ally Omar Bongo had recently been incensed by Obiang and he had an old relationship with Ely Calil, and French interests were jealous of American oil companies in Equatorial Guinea. Yet I have seen no evidence of their involvement, and French interests were hardly likely to use British mercenaries.

Did Obiang stage it to justify a crackdown? Such a baroque international plot seems far beyond his wit. Did the mercenaries act alone? They appear arrogantly to have assumed a God-given right to run the affairs of Africans, and did not expect to be outwitted by them—even after news of their plan had obviously leaked. But Mann's confession states—and others involved back this up[67]—that Mann was approached, not the other way around. If there were other interests, it is possible that the mercenaries may not even know about it.

Johan Smith, a security consultant who knew many of the plotters, said he sent a report in December 2003—four months before the coup—to Michael Westphal, a senior colleague of Donald Rumsfeld, and to two contacts in British intelligence. He sent a second, more detailed, report to the same people in January.[68] A book about the plot, entitled *The Wonga Coup*, details meetings between the plotters and unnamed oil companies, as well as with British and American officials, including other senior staff in the State Department and the Pentagon.[69] Severo Moto also reportedly discussed the plot in advance with the then Spanish prime minister José María Aznar, an old friend of his[70]—although Aznar's office has strongly denied any foreknowledge. Aznar was the third man at the podium in Washington when George W. Bush and Tony Blair launched the "coalition of the willing" ahead of the Iraq war in 2003.

Could the same three countries that led the bungled invasion of Iraq have backed, or tolerated, a second bungled invasion a year later of another oil-rich country run by another brutal dictator, this time involving an arms-length mercenary army?

British, Spanish, and American intelligence services knew about the plot in advance, and did not warn Equatorial Guinea. British foreign secretary

Jack Straw admitted that his government had known about the coup plot since late January 2004—weeks beforehand.[71] "Charges that Britain knew in advance of the plot and failed to warn Obiang's government could not be graver," wrote Britain's *Observer* newspaper. "They suggest that Britain could have acted in contravention of international law; that the government may have provided misleading information about its knowledge; and that Western governments might be turning a blind eye to efforts by some to intervene in the affairs of nation states for commercial gain."[72]

Obiang made tantalizing accusations. "The petroleum wealth of Africa is the new honey that attracts the foreign bees to our home,"[73] he said. "During questioning we have found [the plotters] were financed by enemy powers, by multinational companies," he said. "There are other countries who knew about this attempt and did not contribute information . . . ; we will have to qualify them as enemies. Multinational firms operating here and outside who contributed to this are also enemy companies."[74] Obiang said that Spanish warships had been practicing menacing maneuvers in Equatorial Guinean waters just ahead of the coup, adding that he had been warned of the scheme only by Angola, Zimbabwe, and South Africa. Zimbabwe's home affairs minister said the plotters "were aided by the British Secret Service, that is MI6, American Central Intelligence Agency, and the Spanish secret service."[75] Nick du Toit, the mercenary leader arrested in Malabo, said in court that he had been told that Spain would recognize Moto as leader and that the plot was blessed by unnamed "higher-up politicians" in the United States.[76]

A well-informed contact in Washington added this: "Obiang picked up the information that Riggs had changed its procedures regarding what would happen if the government was overthrown," he said. Previously, Obiang knew that if he was overthrown, his signature meant he could still access the money in Riggs. Soon before the coup, the procedures were changed. "Obiang heard about this. This meant that if there was a coup, the money would go to the coup plotters."

Theories that Spain, the United States and Britain backed, or tolerated, this plot face formidable hurdles. First, the countries have denied it. Second, Zimbabwe's president Robert Mugabe hates the United States and Britain and would gleefully make such allegations, even if false. Third, Mann said his prison testimony was extracted under torture (he later formally withdrew it), so it cannot be trusted. Fourth, if Obiang believed this, he would punish American oil companies, and this has not obviously happened. Fifth, the conspirators had a joint venture with Armengol, Obiang's

brother, suggesting the plot may perhaps have originated lc
nally (when Thatcher was arrested in South Africa, text mes
from a "Military Liberation Committee" to thousands of
subscribers in Equatorial Guinea, claiming that "this attem]
of someone in Obiang's clan.") As for the Riggs money, I
able to confirm my contact's assertion, and even if true, it w
anything.

Perhaps, various sources have suggested to me, the thre
ernments thought that they didn't need to warn Equatoria
they knew the South Africans were already doing so. Or, th;
warn them because rumors of coup plots in Equatorial Gui
penny, or that they had bad intelligence, or that it would l
seen to be warning Equatorial Guinea about this, or that
disorganized.

Finally, the killer question: what could an American or]
be? To topple a nasty dictator? American companies had the
tracts, and would hardly act against the man who guaranteed
This last question is the main reason why many people I h;
doubt that western powers backed the plot.

Yet these objections can all be objected to. The unreliability o
timony is just that: unreliability; it does not prove or dispro
Whether the plot originated inside Equatorial Guinea or no
itself rule out western complicity or support. And maybe Ol
acted against American oil interests because he is intimidated
ing to disrupt the oil industry. Were the British and American
just disorganized? With so much at stake—well over $50 bill
oil—this seems rather unlikely to me.

Now—back to the killer question—what could a Britisl
can motive be?

To find a possible American motive, I will rewind to the
ruling party congress I attended in Bata in 2001. The pres
Teodoro Nguema Obiang, known as Teodorín or TNO, was gr
cotting the proceedings, as he had found out that he was not
made vice president at the congress as he hoped. Javier Espinos
El mundo newspaper tracked Teodorín down and invited me to
him to Teodorín's seaside villa.

Teodorín is the son of President Obiang and his favorite
stancia, and is widely assumed to be Obiang's chosen heir. He

ecord company in Beverly Hills, TNO Entertainment,[77] and a stable of Lamborghinis, Ferraris, and other sports cars in an underground garage in Hollywood. He flies in mechanics to fix his cars that break down at home in Bata and Malabo. A music industry official once described him as "a party guy, who comes in with the fabulous chicks."

At the entrance to his villa, people waited on a bench for an audience with the big man, and chickens and goats roamed by their feet. Bodyguards with short-sleeve suits and walkie-talkies prowled the villa's perimeter. Teodorín, then aged 32, greeted us in a black collarless outfit and sunglasses. Sweeping his hand before him, he gestured to his kidney-shaped swimming pool with an aquamarine mosaic dolphin dancing under the water in the bright sunlight. Behind the pool stood five-foot white alabaster goddesses, and behind them the Atlantic waves rolled in toward his private beach. Unseen over the horizon, frantic oil exploration was gathering pace.

"Do you like my new home?" he asked. "I brought these statues from Italy." He had recently returned from the United States, where, he claimed—improbably—that he had spent three months training with U.S. Special Forces, becoming an honorary captain.

Teodorín owns an airline and large timber concessions that earned him, according to a Riggs Bank internal report, $10 million a year.[78] He did not think this conflicted with his role as minister of forestry. "I am a businessman first and a politician second,"[79] he said with a shrug. He owned property in the United States, once offering $11 million for an 18,000-square-foot duplex on New York's Fifth Avenue, owned by the Saudi arms dealer Adnan Kashoggi. His offer was not accepted, but that much money could feed his nation for a month. He bought a $6 million house in Bel Air, across the street from the actress Farah Fawcett. At the time he was promoting Won G, a Haitian American rapper whose videos for the songs "Nothing's Wrong" and "I love TNO" were crammed with poolside women. Later, he was to have a relationship with the American rapper Eve, whom he showered with expensive cars and other gifts.

Teodorín's Asonga radio station, and his populist style—mixing personal largesse with public grumblings about oil companies—resonate with destitute locals and have helped him build a following among the poor. He funded and ran the Association for the Sons of Obiang, a band of unruly youths whom I saw hanging out in baggy jeans at his villa, playing pool, listening to loud music, and dueling on video arcade games. A diplomat in Malabo said Teodorín had been known to be hysterical and had "never had

to work for anything in his life." Obiang himself has said that his son is "sometimes impulsive."

Teodorín was then feuding with family members, including Obiang's brother Armengol, whom he had punched in 1999. In a recent quarrel the head of national security had sent soldiers to disarm Teodorín's bodyguards. Some of his enemies are army generals linked not to Teodorín's mother Constancia but to Obiang's other wife. "This has been a battle," he mused, stroking impeccably cropped hair. "The conservatives want to hold me back. . . . Practically all of Equatorial Guinea is with me. But this is not a kingdom. There is no king or prince here. There is an understanding now to calm the situation—an understanding reached in a traditional way."

Teodorín had recently issued a bizarre press release. It began by denying a blogger's report that he had been arrested in New York, then railed against the oil companies for not recruiting enough local labor. "There is, however, something far more worrying, and that is the maneuvers of certain individuals . . . I am talking about a certain general. It is important that everyone should know that these maneuvers have resulted in the disarming of my entire security branch, thereby leaving me defenseless both in my own country and abroad. How is it possible? The answer is simple: they are provoked by jealousy of the support I enjoy throughout the country."[80]

Teodorín's foes have kept him out of the oil sector so far—to the relief of oil companies who fear mayhem if this wild, populist, playboy takes over. "Teodorín would wreck the industry," an oil official said. Teodorín was appointed minister of infrastructure in 2003, adding to his forestry portfolio and strengthening his hand. On several occasions he has criticized the oil industry, saying that they have been ripping off Equatorial Guinea. "He wants to be minister of mines and oil. The Americans have refused him," a diplomat said. "He thinks all the family is against him. But the Americans are the biggest problems for him."[81]

This has been, for years, the central question in Equatorial Guinea: Will Teodorín get the oil portfolio, or—worse—take power after Obiang? Here lies a possible American motive for a coup: Obiang is rumored to be ill with prostate cancer, and this hothead, if he took over, might renegotiate the oil contracts and cause chaos in the industry. This might be worth trying to prevent: the year of the plot, Equatorial Guinea only earned 26 percent[82] of the value of its oil for itself, while countries like Angola, operating in equally costly deep waters, typically earn 45 to 50 percent (and will eventually earn much more once costs are paid off; Nigeria earned an average 90 percent from its oil in 2001).[83] The oil companies, it seems, are still earning super-profits in Equatorial Guinea.

Yet this is merely a *possible* American motive, and it does not prove anything. I don't see any big British motive, unless it was to get rid of a notorious dictator.[84]

So who originally cooked up this plot? It wasn't the mercenaries. Was Calil the mastermind, as Mann's torture-induced testimony, backed up by what other plotters said, suggests? Calil strongly denies being involved. Did the American, British and Spanish governments stand out of the way? Were they even more complicit? I don't know. But it sure looks fishy to me.

Equatorial Guinea tried civil suits against the plotters, but the claims were thrown out, and the plotters—apart from those caught *in flagrante* by Africans in Africa—roam free today, untroubled by legal actions. The affair seems to have been dropped, perhaps based on a view that it would have been nice if the plotters had got rid of the brutal dictator. I cannot accept this. These people were not fighting against injustice, but for money. In centuries past they might have been called pirates, and dealt with by the Royal Navy. South Africa, well aware of the dangers of armed destabilization on the super-fragile African continent, has since laudably beefed up its anti-mercenary legislation. Britain and the United States have not bothered. In the twenty-first century, we do not seem to have advanced very far.

Finally, has oil tamed the dictator, and made his people happier and wealthier?

The opposition leader Plácido Micó said the thuggery of years past has abated a bit in Malabo, where expatriates live and where human rights groups monitor most keenly (and lobbyists like Bruce McColm may even have helped persuade Obiang that human rights abuses damage his interests). In the countryside, away from foreign eyes, Plácido reckons things are as bad as ever.

As Equatorial Guinea's per capita income rose from $368 per capita in 1990 to over $2,000 in 2000, the country slipped ten places down the United Nations' Human Development ranking. It now has the dubious distinction of being the country with the greatest negative difference—93 places—between its ranking in terms of human welfare and its income per capita. Agriculture and manufacturing have fallen to less than two percent of GDP between them, while oil claims 93 percent.[85] The share of health and education spending has shrunk.

Obiang said in 2003 that there is no poverty, only "shortages."[86] Yet the IMF in 2005 was gloomier. "Unfortunately," it said, "the country's oil and gas wealth has not yet let to a measurable improvement in living conditions for the majority."[87]

Now, despite all the scandals, powerful Americans line up to praise Obiang. "He came asking for our technical assistance," World Bank president Paul Wolfowitz said in 2005, "so they can manage their newfound wealth, manage it—[applause]—manage it according to the standards of transparency and accountability that will ensure that wealth goes to benefit the people . . . I was very impressed at his leadership."[88] U.S. AID has signed an MOU with Equatorial Guinea that, it says, "will serve as a model for future partnerships." President George W. Bush in May 2006 directed the secretary of defense to begin a new military partnership with Equatorial Guinea.[89]

"You are a good friend and we welcome you," gushed U.S. secretary of state Condoleezza Rice at a press conference for Obiang in Washington in April 2006.[90] (His visit followed George W. Bush's presidential proclamation[91] in 2004 barring corrupt foreign officials from entering the United States.) A spokesman for Obiang was just as effusive. "The United States has no greater friend in central Africa than Equatorial Guinea."[92] Senator Carl Levin took a different view of Ms. Rice's meeting. "The photograph of you and Mr. Obiang will be used by critics of the United States to argue that we are not serious about human rights and democratic reforms."

Back in Equatorial Guinea, just like in Angola's Cabinda province, the American oil workers live isolated from the rest of the country in places like Marathon's Punta Europa complex, a haven carved out of Malabo's chaos. Oil workers call it Pleasantville, and it comes with Wi-Fi internet, direct dialing to the United States, regular power and water supplies, hospitals, and supermarkets. They are building a liquefied natural gas plant nearby, whose gas will be flowing into British homes from late 2007.

"This is the shit-hole of the planet," an American oil worker told a journalist from *Der Spiegel* magazine who visited Equatorial Guinea disguised as a priest.[93] "Our bosses hate the corruption, they hate these guys and most of all they hate the protocol. They're oil titans who have more people working for them than this place has residents. They fly in from Houston in their Lear jets. When they get here, they meet with a minister who decides to cancel the negotiations if anyone dares to sit down before he does or neglects to call him Excelentíssimo . . . The Texans know, of course, that their business partner sitting across the table is no 'Excellency,' but in truth a 'son of a bitch.' Deep in their hearts, they despise themselves. Both sides despise themselves. And each side knows that this is true of the other."

I wonder how much good can come out of a relationship like this. I would not necessarily argue against engaging with Equatorial Guinea, if the aim is to improve matters for ordinary people. Even so, Obiang's boosters take things a bit far, reinforcing a suspicion that any behavior can be ignored in the pursuit of oil. "I think there's a sincere intention on the part of the president," said Marathon Oil's president Clarence Cazalot, "that they really be the model for the way it should be done."[94]

Maybe the oil companies' spin doctors actually believe their own words. "In Equatorial Guinea, ExxonMobil is making a difference," the company said in a statement.[95] "Being a caring neighbor takes many forms. Sometimes it requires muscle as well as financial support. And sometimes it just requires the initiative of one of the world's premier companies, simply it is because it is what caring neighbors do."

8

FRADIQUE DE MENEZES

BATTENING DOWN THE HATCHES

It is hard to build a steady career as an independent writer covering Africa. Oil-rich countries are, because of their oil, expensive to work in, and companies seeking sensitive, hard-to-get information offer fees far above what a journalist might earn. Several journalists I know from Africa's oil zones have moved into commercial intelligence.

Another problem is being watched. Oil-rich African rulers tend not to harm western journalists, since this upsets western governments. But they can afford large secret services; political and economic power are so intertwined that governments need constant intelligence about potential adversaries. Once, as I interviewed a powerful former minister in Luanda, he took the pen from my hand and wrote in my notebook, "This One Is Listening To Us." What disconcerted me most was that the snoop, who might have been of Middle Eastern origin, looked me brazenly in the eye. The point is sometimes just to let people know that someone *might* be listening, to keep everyone nervous and in their place. If this is occasionally an inconvenience for western reporters, it is a much bigger problem for African journalists, who have families on the ground to worry about.

Another difficulty is disenchantment. In the West, the fourth estate helps keep rulers in order. But in sub-Saharan Africa's oil nations, rulers often just ignore the media. All this can be dispiriting. Which is one reason why, in 2001, I got interested in the former Portuguese colony of São Tomé e Principe, a pair of Atlantic islands southwest of Equatorial Guinea's Bioko Island. The tourist guide gave me one incentive to go.

"A veritable paradise on earth awaits the visitor to these remote islands. With cloud-capped volcanoes, jungle greenery, and crystal-clear waters not yet discovered by the skin-diving crowd, São Tomé e Príncipe would be on almost every traveler's list of places to see, if they had only heard of it. The people are the friendliest, the best-educated, and the healthiest by far in central Africa. You'll feel the difference the moment you arrive."[1]

Oil had not been discovered yet, but big oil companies were already circling, sniffing under its sea bed with seismic equipment, and the newly elected president Fradique de Menezes was saying refreshing things about harnessing the oil for the good of his country. Experts in foreign aid were eyeing it too: here was a chance to put in place safeguards ahead of oil production, to try to stave off the disasters that afflict other oil-rich countries nearby. Some were saying that this could become the Monaco of Africa, and Fradique, as he is known, was saying the right things. Might this be the place that finally gets oil right? Or would oil destabilize São Tomé, even before any was found?

I knew I had to meet this unusual president, so I planned a trip for November. On the way, I read an enchanting piece in the *New Yorker*, subtitled "Who Needs Saudi Arabia When You've Got São Tomé?" It described a meeting that President George W. Bush hosted for some African leaders at the Waldorf-Astoria in New York. While other leaders spoke, mostly in French, Bush tapped his pencil and looked bored. But when Fradique stood up and spoke in good English about the common interests of São Tomé and the United States, Bush perked up and the pencil tapping ceased.

Fradique was not talking like the African leaders I am used to: he seemed rather pro-American, and peppered his speeches with buzzwords common in foreign aid circles ("transparency," and "governance"), and with names of African American heroes like James Baldwin and Thurgood Marshall,[2] who may be well known in America but are hardly recognized in Africa. "We should not only ask what America can do for us," he said in one speech, "but what we can do for America."[3]

Fradique also sounded canny about oil. His predecessor had overseen the signing of deeply unfair oil contracts with Mobil (now ExxonMobil) and two smaller companies,[4] and he had brought in big guns to help, including the American lawyer Greg Craig—who has represented the former Soviet dissident Alexander Solzhenitsyn, and President Clinton during the Monica Lewinsky scandal. Craig was introduced to Fradique by Joseph P. Kennedy II, the former Boston congressman who heads Citizens Energy, an unusual company that has mixed charitable giving with pursuits such as trading Angolan and Nigerian oil. The *New Yorker* caught the spirit of a meeting in São Tomé:

> Fradique popped the champagne himself. He chatted about the oil contracts with Greg Craig, who mentioned that he didn't have all the documents he needed. Fradique said that he would have what Craig wanted photocopied straightaway, but the various aides seated around the table began squabbling about who should go to the photocopying machine. "You are the chief of staff, you should go do it," one said. The chief of staff retorted sourly, "Why don't you?" and didn't budge from his chair. In the end, Fradique did the photocopying himself.

The article evoked the image of a well-meaning president who, despite his quirks and faults, was a just leader—pro-American, yet smart enough to see through the charm of people such as the Nigerian president, who once held his hand at a meeting. "They all treat me like I am their sweetheart," Fradique cackled. "I wonder why."

When I arrived, one pleasant surprise followed another. The country consists of two islands—São Tomé, with 140,000 people, and Principe, with 5,000, spread over an area a quarter the size of Rhode Island, or two-thirds of Greater London. Until the fifteenth century, the islands were uninhabited, but Portuguese slavers visited and they became stopping points for slavers sailing from Africa to the New World. Over the centuries the slavers' blood mixed with that of colonists, convicts, and Jewish children forcibly separated from their parents in Portugal and deported to São Tomé.[5] British chocolate manufacturers boycotted São Toméan cocoa in 1909—ostensibly because of the harsh labor conditions, but perhaps because it was competing too strongly against British cocoa interests. "The native of S. Thomé," wrote the British cocoa magnate William Cadbury, "is a brown-skinned individual, insolent, lazy and lawless. His women, some of whom, I am told, are possessed of a certain dusky comeliness, are notoriously loose. If you meet them on a muddy road and there is a dry path only wide enough for one person, you must walk in the mud."[6]

In this melting pot, ethnic divisions dissolved into a society that is more Caribbean than African, and São Tomé does not suffer the ethnic violence that plagues mainland Africa. Still, seven or eight bickering family groups have loomed over São Toméan politics for years,[7] producing interlocking, shifting factions—illustrating that the divisions that divide societies and generate corruption in Africa are not just ethnically based.

Unlike in oil-producing nations that I am used to, where the president is usually all-conquering, power in São Tomé is divided between the president and the government. The last two presidents had left office peacefully after free elections, and parliament has genuine power, rooted in the constitution. It fares better than most sub-Saharan African countries in the World Bank's "governance indicators": voice and accountability, political stability, rule of the law, and the like. Despite the constant quarrelling between unstable factions, São Tomé is generally peaceful, and life expectancy is much higher here than in the oil-rich countries that frame São Tomé to the north and east.[8]

A taxi driver suggested another reason for peace. "If you do someone harm here, where will you run?" he asked. "You will meet your victim, or his family, again." Some locals joke that the initials of São Tomé e Principe—STP—really stand for the Portuguese *somos todos primos,* "we are all cousins." In this tropical goldfish bowl, bitter enemies are often related, and grievances do not last long.

In São Tomé Town the buildings are mostly modest Portuguese-era dwellings, many with red-tiled roofs, interspersed with wide boulevards. Even when it is sunny in town, the volcano top that dominates the island is often obscured by dark clouds. Out of town, there are scrubby villages and huge, overgrown cocoa estates, still hosting raucous communities. Orange crabs roam the roadsides, waving eyes and claws defiantly; giant turtles lay eggs on the beaches at night, and fishing boats sometimes bring in marlin or sailfish. Sit at the Paraíso dos Grelhados, an old blue shipping container under a great sea-front cashew tree, order fresh red snapper with pepper sauce and roasted breadfruit, and lean back—it is easy to imagine that the soothing Atlantic breakers rolling endlessly into the beach are also a reason why the islands are so peaceful.

The top politicians all agreed to see me, and the taxi drivers knew where they all lived—another pleasant shock after the intimidating secrecy of Africa's other oil zones. Fradique's affable press officer, who appeared to be in a loud bar when I called, said that it should be easy enough to meet the president. The ministry dealing with oil was just a large red-tiled house, and an office recently set up for oil matters struck me as being more like a poet's or a musician's residence. When I asked to use the bathroom there, a secretary looked aghast, then rushed away (and cleaned it, I suppose) before letting me in.

The politicians all knew of the *New Yorker* article, perhaps São Tomé's biggest-ever media splash. They mostly hated it, largely because it was quite kind to Fradique, a light-skinned mestiço who, his enemies point out, fought in the colonial wars on the side of the Portuguese. He was born in São Tomé in 1942, studied at private schools in Portugal and Belgium, then worked for Marconi, ITT, Archer Daniels Midland, Goodyear, and Memorex before becoming ambassador in Brussels in 1983. He returned home in 1986, and soon became the country's top cocoa trader and cement importer. He has few family links: his Belgian wife died of malaria in 1993, and he spent years overseas.[9] So he is liberated from some of the African obligations to do favors to close kin, though still subject to huge pressures from São Tomé's factions. Some say his lack of family makes him less stable. "He acts like a wild man at times," one observer said. "At one level he believes in stuff that is right for São Tomé. . . . He is willing to use input from the West where a lot of other leaders are not. But he is hard to fathom. I am not sure even he knows what he wants."

Guilherme Posser da Costa, a former prime minister, ambled over to my hotel clutching a briefcase and papers. His car had broken down, so he had to walk. When I asked about Fradique, he scowled. "He is completely indisciplined," he said. "He wants more power and he does not respect the constitution—he has a rebellious character, and is extremely autocratic. We don't know how to advise him. He is so unpredictable."

Carlos Tiny, a failed presidential candidate, received me with tea and biscuits under dangling forest creepers in his garden. Two cell phones on the table rang every few minutes, though several calls were from a family member giving him grief over a flat tire. Tiny led me on a twisting route through bitter old political quarrels, with fallings-out and reunions that open and close again, like crevasses in a moving glacier. He was not currently in government. "Me, a minister?" he asked. "Any street kid can be a minister here. We have more ex-ministers per capita than anywhere. I am not interested." He was also frank about transparency. "In public everyone agrees to transparency," he said. "In private, it is different."

Tiny is one of Fradique's most vocal antagonists. "Fradique wants to turn this into a banana state. Unless we put things in some sort of order now, this will be a total mess. Everything that happened in those other African countries started like this." My early impression of Fradique as a valiant knight battling political dragons was giving way to a more complex picture. Fradique was less Saint George, some said, more Don Quixote.

Once I had penetrated the byzantine complexity of São Tomé's shifting politics, it was clear that the central problem was an old power struggle between the presidency and the legislative: parliamentarians and government officials wanted constitutional changes in their favor; Fradique wanted the opposite. This debate was stirred up by the question of who would control the oil. All were stressing the need to batten down the hatches as the oil storm approached. "This is a battle for position ahead of oil," the former prime minister Posser da Costa had said wearily. "This accursed oil which is coming: people are not ready."[10]

I also met São Tomé's first president, Manuel Pinto da Costa, who received me in a modest office on the seafront near Marlin Beach. Elegantly aging in his mid-60s, with very dark skin and silver-flecked hair, he chatted about how he got to become president.

In the late 1950s, long before independence, Pinto da Costa fled as an anticolonial activist to Europe, where the Portuguese and French secret services harassed him. He drifted to West Berlin, from where you could easily cross east and get study grants. "I did not speak German and we were also afraid," he said, opening his eyes wide and stiffening his fingers into claws, "of this Communist monster." As East Germans flocked West, Africans were heading the other way. Pinto da Costa followed, and studied at Humboldt University. He rose to the top of São Tomé's small group of politicians in exile.

In 1975, after the right-wing dictatorship fell in Portugal, São Tomé won independence and Pinto da Costa returned home to become, aged only 37, Africa's youngest head of state. "The Portuguese flag went down, ours went up, and I was president. I went to the palace and sat in my office. *E agora?* What now? How do you govern a country?"

Not even five people in the country had university educations, so he helped teach people to read and write. After five centuries of exploitation, he nationalized the plantations and asked for foreign help: North Korea sent advisors, Cuba sent doctors and military trainers, Chairman Mao built a hulking national assembly, Angola sent oil and soldiers, the East Germans refurbished an old brewery, and the Russians set up a radar station to watch this strategic piece of African coastline.

"The world was bipolar: NATO had supported Portugal, so we got help from the other side. They called us communists. But it had nothing to do with communism."

By the mid-1980s cocoa prices had collapsed and, while other African leaders like Angola's President José Eduardo dos Santos clung to Marxism-

Leninism, Pinto da Costa switched sides, hoping that western patrons would be more generous.[11] He signed up with the World Bank and the IMF, and opened São Tomé to private investors. Many of those who came already knew the rough-and-tumble of places like Angola, and profited from São Tomé's openness, even naïveté, about the world. Stories abound of canny Nigerians removing the innards of fridges before selling the shells, at full price, to cheerful locals; much of the islands' tourist potential is locked up in exclusive long-term contracts held by South African, Portuguese, Angolan, and French business figures with colorful pasts. The biggest was Chris Hellinger, a German who made a fortune in Angola's diamond zones. "The big do not teach the small how to negotiate," Pinto da Costa said. "We pay a price for our ignorance."

São Tomé's most salable asset was national sovereignty itself. Schemes sprang up, many benefiting private individuals but not the budget. Foreign ships were registered under São Tomé's flag (the president[12] once had to ask an honorary consul in Amsterdam to ask Lloyds of London how many ships they had registered).[13] A fugitive Spanish terrorist was once appointed ambassador to the United Nations in Geneva)[14]; São Toméan coins that never circulated at home were minted for foreign collectors; and colorful stamps were produced, featuring celebrities like the Beatles or Marilyn Monroe. The Frenchman Charles Pasqua, and others, tried but failed to set up offshore banks,[15] duty free zones, and a deepwater port. There was a general free-for-all in São Tomé long before oil.

Pinto da Costa ceded power after elections in 1991 to a rival, Miguel Trovoada, under whose watch new schemes emerged. Pornographers got hold of the internet ".st" domain, making São Tomé, according to one survey,[16] one of the world's "porniest" countries, measured by the number of porn pages.

The economy festered: in 1995 the treasury got less than $10,000 from customs import duties,[17] while total annual exports (mostly of cocoa sold by Fradique's company) hovered at around $5 million. External debt grew to six times gross domestic product,[18] and foreign aid funded 97 percent of the government's capital spending program.[19] It had become, in the words of three academics, an "unviable state."[20] President Trovoada, despite fierce opposition from his government,[21] officially recognized Taiwan in May 1997, and was rewarded with $10 million a year in Taiwanese aid.

In 1997 a Washington lobbyist, Noreen Wilson, got interested in São Tomé, through some South Africans seeking opportunities in Africa after the fall of apartheid. Noreen's cousin worked for a waste management company, Environmental Remediation Holding Corporation (ERHC), whose wild CEO, Sam Bass, was fond of guns and custom-made sharkskin suits. He once bit off a piece of someone's ear in a fight in Louisiana, then began laying cash on the table to compensate; when they got to a few thousand dollars, the man took it. "Sam actually determined what the price of an ear was," Noreen said. "True story!" Another time Bass divorced his wife, who demanded half of everything in the settlement. So, with the help of a friendly oil crew, Bass cut their marital home in half, leaving her the bit without the bedroom.

Noreen Wilson joined ERHC, and in 1997 she flew to the islands to negotiate. ERHC agreed to pay São Tomé $5 million, do a feasibility study, and help get oil exploration started. In exchange, they agreed to take a big share in a new national oil company, a five percent overriding royalty, plus at least 40 percent of profits from all future oil operations (including anyone else's). They would negotiate with other oil companies on behalf of São Tomé. Consider that several billion barrels of oil might be found there, multiply by a conservative $30 per barrel, and ERHC clearly had a very juicy deal. From then on, things began to go badly for São Tomé.

Gavin Hayman of the campaigning group Global Witness described the contract, along with a couple of others that came later, as "among the least transparent, most egregious deals of all time."

Wilson rejects this, saying—in an echo of comments made by Exxon-Mobil about how African governments should offer generous terms[22]—that the deal was so rich because nobody else was interested in a zone in water depths that were then beyond the reach of world drilling technology, and which lay in disputed maritime territory. "We met with parliament, with the president. With the World Bank. With *everyone*. It was promulgated in law; this was not something we did overnight. They all said they had no problem with the deal. Nobody thought there was any oil, and as long as we were spending money on São Tomé, that was just fine."

"How did I get the five percent? They said, 'OK, what do you want?' I said, 'I tell you what. I sold real estate for a living, getting a ten percent commission on land. I will take five percent on water.' It was that simple, folks. We always understood from day one that there would be, as likely as not, renegotiation."

It was also clear to those in the know that drilling technology was advancing so fast that São Tomé's most prospective deepwater zones would soon be within reach.

I tracked down former President Trovoada, on whose watch the ERHC deal was signed. His walled residence was guarded by two sentries, and the anteroom where he received me boasted two-foot ivory lamp-shades, decorated ostrich eggs, and an ornate three-foot mermaid in gold and rusted copper-green, holding up a glass tabletop with cascading red flowers and orchids. He chided some children in an adjoining room in French, then came through to join me. Unprompted, he was immediately defensive. "There are no secrets here. Get a car in customs, and the next day everyone says, 'he got the most luxurious car in the world!' No, I challenge anyone to demonstrate that I have been corrupt." His enemies, he said, had been opposed to the Taiwan aid deal and signed up with ERHC to prove that there was an alternative to Taiwan's money. "The deal was disastrous," he told me.[23] "It created *confusão*—I watched from a distance. It was nothing to do with me."

ERHC helped set up the national oil company, with Noreen Wilson as president. In September 1998, ERHC flew president Trovoada and several others to the United States, where they were received by Senator John Ashcroft (later, controversially, appointed U.S. attorney general by President George W. Bush) and Jesse Jackson. One insider described a dinner that President Trovoada attended, where Mobil presented the São Toméans with a 50-page contract and told them to take it or leave it. Trovoada sent it to oil industry experts, who told them that it was not industry standard. "Mobil went to the prime minister," he recalls, "and said, 'We will pull out unless you sign.' So they signed."

ERHC's dealings are in the public domain, because it must disclose them under stock exchange rules. But bigger companies like Mobil escape this scrutiny, because their global operations are so large that their tiddly little dealings around São Tomé were not considered material to the share price. Yet São Tomé's society is so open that it was not hard to find the Mobil contract, another staggering deal. Mobil got rights of first refusal on *all* of the 22 huge offshore licenses then being offered by São Tomé—effectively the country's entire potential oil zone, plus options to take areas outside the zone if they wanted that, too—in exchange for a signature bonus of just $5 million—peanuts in industry terms—and if Mobil discovered oil, they would get *100 percent of profits* for Mobil (after just 7 percent royalties) for an operation producing up to 50,000 barrels per day, 90 percent for up to 100,000 barrels per day, and so on.[24] It was a bit like what happened in Equatorial Guinea, where a small American firm broke the ice, then Mobil moved in and got the good contracts ahead of the competition. In São

Tomé tiny ERHC launched itself, rottweilerlike, into the political dog-fights needed to open up the industry, while Mobil stood politely back, then strolled in to the spaces that ERHC had opened.

Yet now Mobil and ERHC both had potentially conflicting claims to some of São Tomé's best oil licenses. If the ERHC deal was when things began to go wrong for São Tomé, then the Mobil deal of 1998 was when things began to go wrong for ERHC. According to Wilson, the newly merged ExxonMobil set out to shake ERHC out of São Tomé, to free up ERHC's rights. "Five minutes after signing the deal [with Mobil]," she said, "they went to the government and said, 'Oh, you don't need ERHC any more, you can go around 'em, we can help you, we will get you the lawyers."

This happened as Mobil was finalizing its merger with Exxon, and the Mobil team with whom ERHC had built good relations was removed. "Everyone we had dealt with—we lost everybody. You didn't really have Mobil in charge any more; you had Exxon people, who have a very, very different mentality."

The ERHC deal began to sour. The U.S. Securities and Exchange Commission began investigating the company for corruption (no wrong-doing was found). "The U.S. State Department, the lawyers, made up their mind that this small little company had done wrong because the big guys had told them we had." ExxonMobil eventually got its rights to take prece-dence over ERHC's rights, where they conflicted.[25]

Indeed, attacks on ERHC since then—though they may have consid-erable justification—have also struck me as surprisingly energetic and high profile. The São Toméans went so far as to instruct the lawyer who de-fended Elian Gonzalez and Bill Clinton in the Monica Lewinsky saga. Later, I visited another lawyer who had worked against ERHC; not only were his offices just two or three minutes' walk from the White House, but to get there I had to find a secretary to come into the elevator to unlock ac-cess to the floor I needed to get to.

ERHC had money troubles, too: world oil prices were flirting with $10 a barrel, and interest in a company with shaky rights in an untested country was low. So, in August 1999, ERHC shareholders sold a controlling inter-est to Talisman, a company based in Little Rock, Arkansas. Talisman's CEO, Geoffrey Tirman, specialized in "distressed securities," or poorly per-forming companies, which could be turned around and sold off at a profit. Tirman's strategy, it seems, was to pressure São Tomé to stick to the deal, to squeeze out more value from the contract. He wrote to the São Toméans, saying that the Clinton administration was "fully supportive of our position" and that he would be arriving shortly with a "very senior-level person from the current U.S. administration. . . . [A]ny attempt to deviate from the agreements currently in place will not be tolerated."[26]

Tirman arrived in São Tomé with a former aide to U.S. President Bill Clinton named Mark Middleton, also (like Clinton and Talisman) from Arkansas. The talks broke down, and at the airport as they left Tirman publicly accused senior São Toméans of asking for bribes. The São Toméans cried outrage. Tirman apologized, but the São Toméans said that they would no longer honor the contract. Tirman filed for arbitration in Paris, and ERHC's share price plunged. A lawyer familiar with this issue said he thought that a court might well have upheld São Tomé's termination, and that São Tomé could have paid back the original $5 million, plus expenses and interest. Not doing that, he said, was their "fatal mistake."

By then, auditors were raising doubts about ERHC's viability as a going concern,[27] partly because São Tomé had a border dispute with Nigeria, which affected their license. This was no small matter, as former president Trovoada explained. "We delineated and agreed on our borders with Gabon and Equatorial Guinea. Then came Nigeria, and—Pumba!" he exclaimed, smacking his hands together. "We hit an obstacle."

ERHC needed a bold new strategy. They found it in a Nigerian businessman, Emeka Offor, who had become highly influential in Nigeria under the brutal former dictator Sáni Abacha.[28] In 2001 Tirman's group sold a controlling interest in ERHC to Offor for $6 million, though the Americans kept a significant share. (São Tomé then also signed another horribly skewed contract with a Norwegian seismic company, PGS.)

In doing this, ERHC changed its spots, becoming one of those African oil companies that effectively trade on their political influence, which they use to nail down their rights, and then bring in foreign oil firms as partners to provide the expertise and to pay the exploration and development costs.

Offor had already been busy, and in February 2001, six days after he bought ERHC, Nigeria and São Tomé settled their dispute.

One way to settle such disputes is to use the principle of equidistance: make the geometric median line between the countries the border; indeed, both São Tomé (and Nigeria since 1978) had accepted this general principle. But Nigeria had recently changed its position, away from median lines to "negotiated settlement."[29] With a powerful military, a population 1,000 times greater than São Tomé's, and long and wily experience in oil, this new strategy played to Nigeria's strengths. Nigeria also wanted to settle fast, before its ally, president Trovoada, handed over power in elections. For Trovoada, bringing in Nigeria had the benefit of marginalizing Angola, which was allied with his political enemies in São Tomé.

The quick solution was this: instead of delineating the border, the two sides set up a Joint Development Zone (JDZ) in the disputed area, overseen by both parties and with revenues split 60:40 in Nigeria's favor, reflecting Nigeria's weight and expertise. The unfair bit is rarely mentioned: except for a tiny notch, the *entire* JDZ lay on São Tomé's side of the median line, which until 1998 both sides might have agreed was the border.[30] São Toméans might argue that with this treaty, Nigeria raided 60 percent of their oil.

Three months after the deal, ERHC and São Tomé settled their arbitration—this was witnessed, curiously, by Nigeria's minister of state for foreign affairs.[31] The new contract was perhaps only a touch less skewed than the original American-designed one: an IMF simulation, using very conservative forecasts, estimated that ERHC, for little more than its initial $5 million down payment in 1997, might earn a cumulative $1.4 billion.[32] Three academics said that they were not aware of a similar precedent in African oil history since the end of colonialism[33] (but Mobil's first contract in Equatorial Guinea, or Elf's contracts in Gabon and Congo, might have given ERHC stiff competition). São Tomé later admitted to the IMF that the secret renegotiation "lacked transparency" and raised "serious governance issues."[34]

ERHC's share price rocketed. The American shareholders saw the rump of shares that they had retained—not including the $6 million Offor paid them for his controlling stake—become worth more than the entire company they sold.

Two months after *that,* Fradique became president. The IMF persuaded São Tomé to send the oil contracts for analysis to the World Bank and the IMF, and to the high-profile lawyers described in the *New Yorker* article. They said that the new contracts were still grossly unfair. So Fradique wrote to ERHC saying the new contracts would be terminated. Though he had no constitutional right to terminate oil contracts, he had set down a political marker. And he had made some powerful new enemies.

At the end of my trip, Fradique's press officer, Guilherme, told me to come immediately to the presidency. He was horrified that I was not wearing a tie, but was reassured when I pulled one from my pocket and put it on. We went in to meet the president.

Fradique is a brawny, ebullient man with a shock of graying frizzy hair, whose voice and personal energy reminded me of the frightening gangster played by Joe Pesci in the movie *Goodfellas*. He wore a shirt with short sleeves revealing a rough, fading tattoo on one arm that was partially obscured but appeared to say something like S.U.B.R. I was too intimidated to ask about it. Fradique started by asking me if I really was who I said I was. It is hard to prove something like that on the spur of the moment, so I felt shifty for the rest of the interview. When I asked him an admittedly ill-informed question, he turned on Guilherme. "How can you bring people like this to me?" he repeated twice, very loudly. Guilherme quailed.

I switched to a more personal note. Who did Fradique admire most, I asked. "Myself!" he answered, straight away. "I am joking, of course!" he guffawed, and settled for Nelson Mandela. The interview settled down.

Fradique said that he had never intended to be president. "When I leave office I will return to my business, and I can live in freedom." I put to him what his enemies said: he wants to be king of oil; he is autocratic, undisciplined, and does not listen. Fradique cut me off angrily before I could finish. "Me? Autocratic? You are being manipulated! When I see the insistence and the aggression . . . I am afraid of this kind of behavior."

One allegation was more troubling. Rumors were circulating that the Nigerian businessman Offor, who had bought ERHC, had paid $100,000 to one of Fradique's companies in February 2002, just ahead of some legislative elections. The key to political success in such countries depends partly on your position in interlocking webs of blackmail and intrigue, where politicians routinely collect dirt about each other, as a source of future leverage. This can generate a vicious circular dynamic: the politicians that rulers allow to rise will be those with the most embarrassing skeletons, who will therefore be easiest to control. The clean politicians, over whom there is less of this leverage, are often not promoted. Antony Goldman, a London-based analyst, explains that while this is part of politics everywhere, in the oil-rich Gulf of Guinea it is central to the business of government. "This campaign payment to Fradique exactly fits this pattern."

Fradique admitted having received the payment, but said it was one half of a legitimate two-way foreign exchange transaction.[35] "ERHC sent me a letter asking me if I had the authority to renegotiate," he began. "Then suddenly this thing—this $100,000—appears on the Internet. They sent this to someone—maybe to pressure me. I don't know. They probably think that by doing this I will forget the renegotiation. By this stupid and silly thing, they think that they will minimize Fradique de Menezes. No. Now I will be stronger than ever." (Later, however, he changed his story and accepted that this money was for his election campaign.)

I asked Patrice Trovoada, the son of the former president Miguel Trovoada, about his links with Fradique. Patrice is a slightly chubby, wealthy young man with a Mona Lisa smirk, and he wears the sharp tailoring more styled to French-speaking Gabon than to informal São Tomé. He drives around in a Hummer with bodyguards, according to one resident, "dressed like members of an L.A. street gang." He jets between Houston, Paris, Abidjan, and São Tomé, says he owns a construction company in Texas, and reels off a list of friends and business associates including Libya's Muammar Qadaffi and Russia's Gazprom, and stretching to South America, India, Taiwan, and mainland China.

In 2001, he said, his father's two-term presidential mandate was expiring, and they wanted a malleable replacement. "We chose Fradique from scratch," he said. "We had to choose someone. We had originally thought about (another) guy . . . but he had no balls: 'I want it—I don't want it,'" Patrice said, swaying his hands back and forth across the tabletop. "So we found this guy Fradique—his wife had passed away, he was close to our youth—and he loves parties. We did everything—we wrote his program, and Emeka Offor paid to support his campaign."

At the time of my interview with Fradique, he had been grousing at Nigeria and ERHC, and Nigeria's president, Olusegun Obasanjo, had just written him a rather menacing letter telling him to pipe down, and suggesting that Nigerian soldiers might hold exercises on the islands. Fradique was furious. "They [the Nigerians] said São Tomé must close the mouth. This note—I did not like it at all. Meddling in our internal affairs! I will not allow it. They have found somebody who likes to talk. The president will be like this until he dies."

So, intimidated by Nigeria, Fradique had been seeking new friends. The Institute for Advanced Strategic and Political Studies (IASPS), an Israeli-American group concerned (among other things) about U.S. reliance on Arab oil, had in 2002 floated the idea of an American military regional home port on São Tomé, to help guard the oil supplies from West Africa. "We have no patrol boats, no coast guard, no patrol planes, and we are a small and vulnerable island state," Fradique said. "If the United States or anyone wants to protect us, it would be welcome." The idea of a military base never really got off the ground in Washington, however, but men with crew cuts and American accents have since been spotted on the islands.

With the IMF's help, Fradique pressed on in his campaign against the oil firms, and in early 2003, ExxonMobil agreed to cut its rights back to 40 percent of one license, plus 25 percent of two more. (By then, the company had enough geological information to know which the best blocks were, so although ExxonMobil ceded a lot on paper, its concession was probably far less than it might seem.) A new deal was also negotiated with ERHC.[36] Yet even after this, the losses to São Tomé from ERHC's special terms were still estimated at an absolute minimum of $60 million[37]—worth roughly ten times São Tomé's total exports. This deal was signed just days before the presidential election that returned Nigeria's President Obasanjo to power. One insider said the São Toméans were told to sign immediately or someone "worse than Obasanjo" would take over, or the president would be too busy to think about São Tomé for a long time. Not for the first time, the Nigerians (who could afford to wait, as they had so much oil already) used the threat of delay to get São Tomé to concede far more than they might have. It was also odd that Fradique, after all his arguing, turned down offers of a credit line from the World Bank and pro bono assistance from an oil contract expert to help renegotiate the ERHC contract. Sixty prominent São Toméans subsequently wrote an open letter accusing him of a lack of transparency in the oil dealings.[38]

The new deal was accompanied by new instability. The same week, days before the entire Joint Development Zone was to be tendered to international bidding, police shot a man dead in a riot in front of the presidential palace. It was the first time São Tomé's police had killed a demonstrator since independence. Not long after that, new agitators turned up: some São Toméan adventurers who had fought as mercenaries[39] overseas (and who had launched a botched coup attempt in São Tomé in 1988). With serious combat experience in Angola, Sierra Leone, Papua New Guinea, and Congo, these men made people nervous. They started threatening new demonstrations. Then, in July, they staged a bloodless coup while Fradique was in Nigeria, attending the U.S.-backed Leon H. Sullivan meeting.

Nigeria's president Obasanjo cooked up a caucus of regional leaders, who flew to São Tomé and pressured the plotters to hand power back to Fradique. After that, Fradique had to accept Patrice Trovoada, his scheming political adversary who was by then a Nigerian ally, as his oil advisor. And, true to the forgiving spirit of São Tomé's incestuous politics, the plotters were pardoned. I later caught up with one of them, a wild-eyed man called Sabino Santos. He was sipping coffee nonchalantly in a seafront café by the Miramar Hotel. He was vague about why they mounted the coup, citing hazy historical grievances. "Oil was at the bottom of this. Yes. It's oil," he said, railing at corrupt politicians gorging on oil money. Yet he denied that any particular oil interests had backed his men. Later, I prodded Fradique's friend, the former foreign minister Mateus Meira Rita,[40] about who was behind it. "Oh come on," he said. "You know who did it. Of course you know." But he would say no more. Different interests in São Tomé told me different things. The plotters visited Fradique two days before the coup, and his enemies say he was in on it. Others suspect his local political rivals, or Nigeria. "The Nigerians wanted Fradique to be eternally grateful," an associate of his said, "and to say, 'here is a shot across your bow. Watch it—we can get rid of you any time.'" Odd movements were also reported near the house of former president Trovoada. All sides reject the allegations, each giving half-plausible explanations.

I am still unsure of the *real* story. But Fradique's early public criticisms of Nigeria have since morphed into grumpy camaraderie, and even praise.

The bidding for the offshore licenses, temporarily interrupted by the coup, got underway, in stages, each of which generated destabilizing new controversies.

ERHC and ExxonMobil already had large shares in the best licenses through their previous deals, meaning that bidders would have to partner with them whether they liked it or not. ERHC's presence in particular deterred many companies, and in the end only one or two reputable ones turned up, plus many small Nigerian outfits with little oil expertise (one operated out of a shed) that were just vehicles for Nigerian politicians to win cheap stakes in the oil industry. Except for one from Chevron for the best license where ExxonMobil already had a 40 percent stake prebooked, the bids were so poor that they were rejected. In November 2004, a second bidding round was launched. Oil prices were higher, and this time the American companies Noble, Devon, Anadarko, and Pioneer bid. ERHC not only took its preexisting rights in these licenses, but partnered with the American companies to get even more.

Even though nobody had yet discovered *any* oil, the bidding heated up politics in São Tomé to the boiling point. Other well-connected Nigerian firms, some of which partnered with the new bidders, sent political raiding parties into São Tomé to influence its politicians in their favor, stirring up the politics once again. One specially constituted company was attached to a favored consortium, then had its shares handed out to key São Toméans.

Accusations flew like arrows between Fradique's camp and those of his adversaries; at one point Fradique remarked that it was like being a spectator at a tennis match. He threatened to dissolve parliament, calling its raucous members "a bunch of delinquents" or "white collar criminals." A former prime minister got a three-month suspended jail term after attacking and smashing up the office of the attorney general, one of Fradique's allies.[41] Fradique was jumpy: he had his guards chase down a lad from a poor slum who shouted at his passing motorcade that he wanted some oil money, then personally grilled the hapless youth at his residence. It was not serious: he quickly freed the youngster who, apart from ringing ears, was unharmed. As usual, oil was not the direct cause of most of these incidents, but oil fever had fostered a pervasive antsy mood.

In early 2005, Chevron and ExxonMobil made the $123 million down payment for their rights, split 60-40 in Nigeria's favor under the rules of the shared Joint Development Zone, yielding São Tomé nearly $50 million. But of course things were not so simple. The Joint Development Authority had curiously nominated a Nigerian bank, owned by a close political ally of Nigerian President Obasanjo, to receive the cash. And the bank refused to release the funds to São Tomé. The Nigerians were saying, sotto voce, that they would not release the Chevron cash until São Tomé agreed to the Nigerian choice of companies in the oil bidding round. A civil service strike was being prepared in São Tomé just then, and the strikers were asking—amid feverish talk of imminent oil money—that the minimum wage be nearly tripled.

São Tomé was under tremendous pressure to get the money fast. Nigerian President Obasanjo flew in, collected the politicians at the presidency, and told them they had effectively no choice but to accept the Nigerian choices in the bidding round. I was outside in the courtyard, and watched Obasanjo stroll down the red carpet, robes fluttering in the hot afternoon breeze. He muttered a few gruff, inconsequential words to assembled reporters, got in his plane, and flew home. São Tomé capitulated to the Nigerian choice of bids, and the Nigerian bank released Chevron's cash. (This was lucky timing for São Tomé: the bank collapsed a few weeks later after a Nigerian fraud investigation.)

Then the government collapsed. The prime minister resigned, citing irregularities in the oil bidding. It was São Tomé's sixth change of government in three years.

The turmoil did not end there. Some American companies whose bids had been accepted withdrew in unclear circumstances, and the attorney general, helped by high-profile American lawyers, produced a report that attacked ERHC's contracts as "extraordinarily one sided," and saying that the companies may have made improper payments to government officials,[42] recommended a referral to the U.S. Justice Department in view of possible violations of the U.S. Foreign Corrupt Practices Act.[43] Texas officials used a search warrant to search ERHC premises again in May 2006, looking for evidence of ERHC's links with São Tomé and Nigeria, and the investigation is proceeding.[44]

Nigerian officials deny forcing anything on São Tomé, arguing that the problems stem from São Tomé's messy politics. ERHC's shareholders, for their part, cannot hide their exasperation at São Tomé's constant changes of heart. "That is why oil companies prefer dealing with Nigeria," Noreen Wilson said. "It doesn't matter if Nigeria has a coup, or who kills who tomorrow: oil flows and they don't screw with oil deals. São Tomé could have had oil years ago if only they had stood by a deal they made."

And so the process limps onward. Though oil licenses are now allocated, big decisions are constantly needed in this complex industry, needing both São Toméan and Nigerian approval. To get what they want at each stage, Nigerians may have to keep meddling in São Toméan politics, generating more instability. I cannot help wondering if São Toméans might have been better off if all the decision-making authority had been handed to Nigeria, perhaps in exchange for more of the proceeds. This way they might, paradoxically, have retained more independence from Nigeria.

What of Fradique? Has, or was, he turned? "We are bombarded with seminars, conferences, and lectures on transparency and good governance," a friend of his quotes him as saying. "But when the Nigerians threaten, who will call Obasanjo and tell him to lay off? Tony Blair? George Bush? No! Because [of Nigeria's oil!] Will Transparency International, or Global Witness, ring up? No! Those transparency NGOS have no front-end mechanisms to help you avoid corruption. They just put you on the list after you've done it. They are not aid programs, they are just criticism programs. Meanwhile, my neck is on the line. I am calling out for help, and nobody is coming."

Yet an image of him as a noble warrior fighting against political dragons and Big Oil is plainly misleading, too. When the attorney general said in 2005 that São Tomé should still consider canceling ERHC's deal, Fradique gave Reuters a wistful interview, saying the contract could not be altered because Nigeria had to agree to any changes, too. "There is nothing to be done now." He was reelected as president in July 2006, with Nigerian help.

More than half of ERHC's shares are now held by Americans, following successive cash injections that have kept the embattled company alive, but diluted everyone's shareholdings. Internet bulletin boards are filled with slavering opinions about ERHC stock, written by what one of ERHC's fiercest critics calls "redneck idiots who thought they would be millionaires."

The journalist Ken Silverstein, who wrote several articles excoriating ERHC, said his stories didn't quite have the impact he expected. "I smugly sat back and waited for the fallout, imagining that the taint of scandal surrounding the deal would provoke popular outrage and possibly even an investigation," he wrote. "In place of outrage there was exuberance; the story spread across the Internet, prompting stock speculators to snatch up ERHC shares in hopes of cashing in on São Tomé's misfortune." The journalist took a call from a big investor in an American company that had oil rights in another African country. His reason for calling appeared to be to say that if Silverstein thought ERHC had screwed São Tomé, he should see what his company had done in that other country. "It was clearly his fond hope that I would write a story detailing the whole sordid affair, and thereby trigger a similar run on his firm's share price."[45]

One can get a flavor of investors' São Tomé feeding frenzy by typing the stock symbol ERHE into Ragingbull.com, a site where, in the words of a rival blog, "pumpers, dumpers, bashers, and mashers spin their psychowares day in and day out." Here is a random sample, on a day when ERHC stock stood at about 35 cents:

—Blog boy says "events are moving away from erhe"—anyone got a magic 8 ball or decoder ring????
—SURE thing 3.67 after award u idiot, ROLL the dice.
—RUN QUICK, Offor (taking) all da oil for hisself, or maybe he's a good guy and won't screw us.
—$6 by march goronteed!!! oops. zero now 'cause offor gonna (take) it awl!
—Quick, somebody get me eye of newt and wind of dragon—I feel the need for some witches brew!

This instability happened not only before any oil was found, but also in spite of an oil revenue management law for São Tomé, drawn up with the help of a team from Columbia University and signed in 2004.[46] The law means that most of São Tomé's oil revenues are not to be spent immediately but instead held in a savings fund at the U.S. Federal Reserve, to avoid a sudden, destabilizing rush of oil dollars into the economy. It also envisages better transparency. The fund is supposed to grow into a big offshore pot of cash—a big, permanent temptation for São Tomé's crafty politicians. What is more, the law targets the oil revenues only *after* they have exited the maw of the Joint Nigeria / São Tomé Development Zone. It cannot touch the shenanigans inside the zone. Given the turbulence that the mere promise of oil has brought to São Tomé so far, and the poor record of savings funds[47] in nearby Chad and elsewhere, it is hard to be too optimistic.

Perhaps the former president Pinto da Costa got it right. "When there is a smell of oil, minds get stirred up," he said. "It creates a mirage in people's heads. If we do not know how to manage it, it will be hell here."

If oil is found, if the savings fund survives politicians' appetites, and if Nigeria and the United States cease their meddling, São Tomé's politicians will still need to agree on how to share the cash among themselves. That could be when the fireworks really start.

9

ARCADI GAYDAMAK

BETWEEN GLOBAL BORDERS

In 2000 a western diplomat e-mailed me about some big, subterranean changes in Angola's diamond industry, which produces about a billion dollars a year of high-quality gems. Its unruly northeastern provinces are considered among the world's last great underexplored diamond frontiers. The diplomat's e-mail, reflecting the murkiness of the diamond industry, ended with a single word: "Yuck."

It was once again a time of war. The Lusaka peace agreement and its aftermath, which I had observed for Reuters in 1994, had collapsed after four shaky years, and UNITA rebels were, once more, swarming across the countryside, boosted by billions of dollars earned from diamonds that they had mined during peacetime. Meanwhile, ever more oil was being pumped off the coast, and oil companies had recently paid about a billion dollars in "signature bonuses" for deepwater exploration rights, much of which the government had used to buy arms.[1] This, as a result, was a very hot war, and some very heavy weapons were being deployed. The changes in the diamond industry were billed as a way to crack down on UNITA's diamonds.

A central player in the diamond shake-up was a Moscow-born businessman, Arcadi Gaydamak. He was not especially well known in the West then, but occasional stories were linking him to prominent characters in post-Soviet Russia, where the new capitalist economy looked to many outsiders a bit like a vast underworld. Using multiple nationalities—he has French, Israeli, Angolan, and Canadian (but not Russian) passports—Gaydamak has built up a fortune trading in a range of countries, from Kazakhstan to Angola to Venezuela. He was once quoted as saying he was one of the five richest people in Israel.[2] In one published photograph, he is standing in a room with a secretary and three men in suits; on a table in the foreground is a pile of cash big enough to fill a couple of sports bags. Gaydamak has his arms folded, and he is smirking.[3] He received the Legion d'Honneur in France for his role in a 1995 operation to rescue two French pilots shot down by the Serbs over Bosnia, whose release he secured with the help of Russian president Boris Yeltsin and the Russian secret services. He also helped spring four French "aid workers" (Gaydamak says they were intelligence officers) from captivity in Dagestan two years later. Late in 2000, eleven months after the diplomat's e-mail, French justice issued an international arrest warrant for him when he did not turn up to a summons to answer questions about his role as a financier in the "Angolagate" oil deals in the early 1990s that had enabled the government to buy the arms that turned the tide against UNITA. Later, Gaydamak was probed for his role in an unusual subsequent oil-backed deal to reschedule $5 billion of Angola's foreign debt. With huge influence in Angola's diamond industry—he was giving de Beers big headaches in Angola at the time—and deep involvement in the vast, byzantine world of Angolan oil finance, Gaydamak had also become a gatekeeper into Angola for foreign investors.

In 2005, I set out to track him down. I flew to Angola and visited one of his farms, a project called Terra Verde near Luanda that provides eggs and vegetables to consumers in the capital. To get there I had to drive north through the Petrangol district, named after a local oil refinery. The road is a hot, busy chaos of diesel-belching trucks, beggars, cinder-block dwellings, shipping containers, dirty burlap tents, and piles of rotting or smoking roadside material, much of it covered in reddish brown dust. By the road there were baobab trees with white-painted trunks, and policemen in blue uniforms standing on high ground with AK–47 rifles. A Jehovah's Witness compound lay behind a high wall topped with razor wire; men in tight T-shirts sat nearby on car hoods, chatting, and diesel-gray children scampered among smoking rubbish, throwing stones at rats.

Soon the view opened out onto the giant Cacuaco suburb and dry, rolling hills; to the left of the road I could see salt pans, and blue sea a mile or two away. Blue and white taxis raced up and down this highway at insane speeds; at the nub of one traffic jam, a light truck lay mashed in the road and a dead youth in a green T-shirt hung awkwardly from a window, facing the sky with his arms flung back, dripping blood onto the tarmac. Immediately past the accident, cars were accelerating and driving like maniacs again.

The city turned to bare grassy countryside, and after 15 or 20 minutes we reached Terra Verde. It is a tidy, pleasant-looking compound, flanked by purple bougainvillea bushes, with neatly trimmed lawns, sprinklers, and white flags, along with the red-and-black flags of Angola's ruling MPLA party. Diggers outside were shifting some dark red earth, and inside the perimeter I could see a multistoried stack of chickens, and low upturned half-pipe greenhouses behind big warehouses. A guard at the gate was wary of me but he let me go in and find the manager, a young Israeli in a white shirt and with a thick accent. He gave me the phone number of one of Gaydamak's managers in Angola, and I headed back to Luanda.

On the return trip the body was laid on a blanket and a fat woman in a shawl was crouched over him, stroking his bloodied head and singing or weeping. We passed a new monument to the Heroes of Kifangondo, commemorating an old army victory. A band was playing, and the monument swarmed with armed police and expensive sports utility vehicles, including a Hummer. Confident men in sharp suits were milling about with their bodyguards for the monument's inauguration. My driver was jittery about approaching it, so we drove on. Back in Luanda, I called Gaydamak's associate to ask for an interview. The Man, as some people in Luanda know him, was in town, but too busy to see me. Instead, I was invited to visit him in Moscow. Four months later I phoned him, arranged an interview, and took the Malev flight from Amsterdam.

I was nervous: he was the subject of an international arrest warrant, and most stories about him in the media were negative. Gaydamak had successfully sued journalists and publications for making what he considered wild allegations about him. Who knows, I wondered as I landed in Moscow, what this man might be capable of?

Gaydamak kept me waiting for 24 hours; it was "a crazy day," his secretary said. So I did some sightseeing around Red Square, where I bought one of those wooden Matryoshka dolls, with Vladimir Putin on the outside, then Yeltsin, Gorbachev, Brezhnev, and finally Khrushchev nested inside. The following morning, I got the phone call: Gaydamak would see me straightaway. I took a ramshackle Moscow taxi to a nondescript office near the beautiful Kropotkinskaja cathedral, not so far from the Kremlin. A tall, thin young man in a suit ushered me straight through.

Gaydamak is of average height, fairly well built, and fit (he does two hours of kung fu each day); women to whom I showed the one photograph he let me take of him judge him attractive, and younger looking than his 53 years. He gave me a card that calls him "President of the Congress of Jewish Religious Communities and Organisations of Russia."

He received me in a room opening onto a larger room, where some young men in shirts and ties were talking on telephones. Gaydamak was in a suit and stood behind a large tidy desk surrounded by bare walls, against which leaned some rather anodyne classically styled drawings, ready to be mounted on the freshly painted walls. Before I turned on my recording equipment, he made me sign a piece of paper, torn from my notebook, promising to send all his quotes back to him before publication. I had little choice but to agree. He asked to speak in French, but I said my English was better so he switched instantly, speaking clearly with a thick, ponderous Russian accent, making a few mistakes (which I have carefully ironed out here). Gaydamak leaned back and watched me, jiggling gently in a flexible chair. I felt as if he were sizing me up; it was a feeling that would stay with me all through the six hours he was to spend talking to me.

I opened with two big questions: his net worth and the secret of his success. He did not want to answer (an Israeli magazine once said $800 million; others say three times more). I asked whom he admired, but he said this was a question for teenagers, adding, flatly: "I am not a teenager."

His cell phone rang. He spoke in Russian, then in French, then returned to me, smiling. "Good business!" It related to a Russian oil and condensate field he had bought (or bought into): potentially one of the world's ten biggest, he said. He would not say which field, but that day big news was emerging about the giant Shtokman field in the Russian Arctic, controlled by Gazprom. Gaydamak nodded when I asked if he had a stake in Gazprom. How big? "I am huge. Well, huge as an individual, yes. Not huge." But he didn't want to talk more about that. We moved on.

Gaydamak's initial reserve soon evaporated; it was clear that he wanted to tell a long, elaborate story about himself. It was an interview riddled with pet peeves: the transparency campaigners Global Witness, who regularly criticize him; the hypocrisy of western governments; a news story that, he said, twisted some facts about a banquet that Gaydamak once attended to honor the former New York City mayor Rudolph Giuliani, after Gaydamak had donated a million dollars to a fund for the victims of the 2001 terrorist attacks in the United States. Yet, ultimately, his point boiled down to this: his arrest warrant is unjust: the fruit of conspiracies cooked up by senior French politicians.

Before going into that—though it was often hard to interrupt his powerful flow—I asked how he became so influential in Angola. Gatekeepers like Gaydamak are not uncommon in Africa; such people gain huge influence in weak countries, and become channels through which foreign investors must pass in order to get in. They can get very rich in the process. This influence tends to be self-reinforcing: money begets power, and more power begets more money. Tiny Rowland, the buccaneering late former boss of the African mining conglomerate Lonrho (whom the British prime minister Edward Heath once dubbed "the unpleasant and unacceptable face of capitalism"), was one such gatekeeper; another smaller one was the German-born Christian Hellinger, who sewed up the Angolan diamond industry in the 1980s, then moved and all but took over the business sector of nearby São Tomé. Both Lonhro and Hellinger were very charismatic. Gaydamak, I sensed, was the same. Sometimes the gatekeeper is not a person, but an organization. Elf assumed that role in French-speaking Africa.

I met a gatekeeper once, who spoke to me on condition that he stay anonymous, and I asked how he got so powerful. His reply (after plenty of wine) was immediate. "Charisma, corruption, and perseverance!" he said. Then, after a pause, "and grasping opportunities." Gaydamak's reply to the same question was more subtle and elaborate.

In 1972, aged 20, he was an early Jewish emigrant from the Soviet Union, according to news reports. He went first to Israel, where he met Prime Minister Golda Meir, but he grew disillusioned and left. After a stint as a sailor on an oil tanker he arrived in France with barely a penny in his pocket. "I slept in the park, on a bench," he reminisced. "When you have no roof, you can see only the sky. True, or not?" he asked. "When you start a bit above the bottom, you look in front of you, and a little bit higher, maybe, from time to time. But when you are really at the bottom, you look only at the very top." He seemed pleased with this image: a metaphor, perhaps, for the origin of his huge ambition. (He can appear self-important, too. Once, when he described one of his personality traits—about avoiding rigid plans and being flexible about the future—I said I felt I was similar, and he mildly rebuked me; it seemed he was irritated by my audacity in comparing myself to him: "Am I just saying banalities?")

He enlisted as a bricklayer, then moved into real estate. Next he moved into translation—mostly between French and Russian—and he began meeting the big Soviet trade delegations that rumbled periodically through Paris. As the Soviet system began to crumble in the late 1980s, and as the mad reordering of wealth there began, Gaydamak was one of few people who knew, and spoke the languages of, both sides.

It opened huge opportunities. While it is possible to make money *inside* countries, it is the interstices *between* countries—such as the nexus between France and the fast-changing Russia—where the real money often lies. These in-between worlds are deep, opaque seas of globalized money that are barely troubled by the wills of democratically elected populations, and where there is no effective global policeman. They are zones of great freedom where tax dodgers and African rulers looting their treasuries can use the same mechanisms as Gaydamak and the world's biggest and most reputable companies to mask their activities. It is the secret terrain of legal limbo that so troubled Eva Joly.

"What does 'translator' mean?" Gaydamak asked. "'Translator' means go-between. If you are active in electronics, your position in the business world is usually with people in electronics. If you are a banker, you have relationships with bankers. If you are a politician, you have relationships with politicians. But when you are a translator—a go-between—you know everybody. Through you, people express everything. . . . The majority of translators just use linguistic tools—but that is not enough—you should use the meaning," he said. "I saw what people want. So I became a good translator. I translate the mind. Language doesn't matter. . . . I can explain to a Japanese or a Chinese person how to buy or to sell, or introduce into his head what I want. It is called spiritual pressure. Language was just a tool. I became rich, and I had one of the world's best relational networks, because I worked with ministers, with the heads of big companies. When I arrived back in Russia for the first time"—this was in 1987—"I was one of the successful businesspeople coming from abroad."

He is also, he said, unusually old for a rich Russian. "Here it is not normal to be wealthy when you are old—when the changes came you were already aged 40—you already had your occupation, your problems, etc. Only young people—aged 20—became wealthy," he said. "If you go to meet the president of a bank—if he is over 35, it means something is wrong." This anomaly stems from his unusual international perspective.

In 1992 Gaydamak spotted a new opportunity in Africa, and grasped it immediately. "When you plan something, you think you know the next step—and then the next step. But this is completely illogical. You can never know what the next step is. I never plan—I am, a priori, open; I say yes, and then I see what the situation is," he said. "Planning means stagnation. That is why the planned economic system of the Soviet Union stagnated. Maybe, for these kinds of reasons, I am different from many people. I did it differently because I have a different character."

This was just before I was appointed as Reuters' Angola correspondent. UNITA was rampant across the country: surrounding and shelling terrified populations cooped up in government garrison towns like Kuito. Angola was producing about half a million barrels of oil per day then, mostly offshore, but much of this was mortgaged to repay past loans. Angola needed money for guns, fast.

In April 1993, President dos Santos called an old friend in Paris named Jean-Bernard Curial, a former left-wing idealist who had run Southern African affairs for the French Socialist Party in the 1980s and later oversaw a French foreign ministry scheme to pass food and pharmaceuticals to places like Angola. Curial left immediately for Luanda where President dos Santos, hunched over a map of his country, laid out the problem. "Savimbi," he said, "is coming to Luanda to cut my throat."[5] Curial flew back to Paris, but his task was not going to be easy: French politics was undergoing an uncomfortable "cohabitation" between the Socialist president, François Mitterrand, and a conservative prime minister, Edouard Balladur. Many politicians were old friends of UNITA, and they were blocking any official support for Angola. Yet one man offered to help: Jean-Christophe Mitterrand, the son of the French president. Jean-Christophe no longer ran the Africa cell at the presidency (he was known scornfully as *Papa-m'a-dit*, literally, "Daddy-told-me"), but he still had influence. Old French networks swung into action.[6]

One of Jean-Christophe Mitterrand's contacts was Pierre Falcone, a business partner of Gaydamak's who was well connected in the right-wing French circles that were so active in Congo and Gabon. Many of them were worried about a slow shift in U.S. sympathies away from their Cold War ally UNITA toward the Angolan government, meaning stiffer competition for French companies in Angola's oil industry. The Angolans needed helicopters but could not pay; Gaydamak had some transport helicopters at hand and offered them to the Angolan president, on easy terms. "I gave them a credit—it was unusual at the time. Then I explained my vision of how we should continue."

He and the French networks also prepared an oil-backed loan to help Angola borrow money for the war. An initial $47 million was provided in

1993, and this was followed by another loan, ten times larger, in April 1994, which was used to buy Bulgarian, Russian, and Ukrainian weapons.[7] He said he also brought in Russian advisors to shake up and help equip the Angolan army. These deals were secret: while reporting for Reuters from Angola at the time, I was unaware of them. They were apparently effective—seven months after the second deal, UNITA sued for peace. Later, French magistrates said these deals broke a French arms prohibition on Angola (Gaydamak says the magistrates relied on forged documents). The investigations became a new judicial dossier separate from the Elf affair, and the media dubbed it Angolagate.

Many people believe that the Angolagate deals explain how Gaydamak became a gatekeeper in Angola. He offers a different perspective.

"For years and years . . . the main backer of Angola was the Soviet Union," he said. "When the Soviet Union disappeared, the consequences were huge for Angola. Once the changes came, the new Russia did not care about materially supporting Soviet ideological penetration around the world, and particularly in Angola. So if [the Angolans] wanted something, who would pay? Once the Soviet Union disappeared, it was very, very important to manage the relationship with the new Russia. And that is what I provided. I began to be an intermediary. Russia was changing so quickly, everything was new: you should know where to go, how to go, how to organize. I was the so-called organizer of everything," he said, then he added, with his characteristic hype: "If I had not come in 1993 to organize everything, the [Angolan] government could have collapsed."

Gaydamak's friendly stance toward the Angolan leadership stood at odds with that of many western actors who have often sought to pressure Angola's leaders or otherwise meddle: Ronald Reagan's military support for UNITA in the 1980s, western campaigners shouting about human rights and corruption, critical IMF reports, and French judges dragging everyone's name through the mud. In explaining his different approach, Gaydamak also revealed a contempt for what he called "so-called humanitarian organizations" in the West. "You saw how nice my farm [Terra Verde] is? I am sure it is the world's best agricultural project," he said. "The food—it is the world's best quality. Vegetables and chickens, with nice packing. People should eat with dignity. It is very important for them not to have their food thrown off trucks as the so-called humanitarian organizations do. You see it on CNN: from trucks, or from helicopters, they throw the food down and people fight for it, like dogs."

His behavior, by contrast, he said, was exemplary. "Let us talk about arms. Just to be in arms—it is synonymous with death. Kill. . . . But in this case we stopped the war—and not only in Angola. The reason for the wars in neighboring countries was the war in Angola. And now, no more wars in Africa, in this area of Africa. Let's speak about Angola. For dozens of years it was a heavy, bloody war." he said. He got up from his desk and was pacing. "There were two belligerents. Why did nobody talk about the arms from the other [UNITA] side? The other side's army was completely illegal, not only according to Angolan legislation, but also because they breached international legislation. The United Nations put an embargo on any economic and financial relationship with UNITA. Why did nobody pose the question, How was UNITA's illegal army supplied on a large scale, and who financed it? Why? Strange!" he said, before hinting darkly that it was, once again, French interests. I reminded him that UNITA's diamond-fueled arms trade was also excoriated and was even tackled with a western campaign against "blood diamonds," but he waved this away.

"The legal government of Angola, through the Angolan ministry of defense, bought arms from the very legal Russian government, respecting all international legislation. . . . I was an oil trader," he went on. "Then I convinced international banks to advance the money—around $500 or $600 million—mortgaged against Angolan oil, and this money was paid to the Russian government to supply the arms. My role was to organize and control this financing, to ensure it was spent in a rational way. That is how, very fast, the government stopped the war. That was my involvement."

His points are not entirely without merit, though I consider his regional analysis to be oversimplified and exaggerated.[8]

We moved upstairs from his office for lunch in Gaydamak's rooftop Casual restaurant ("the best French restaurant in the city"). Over exquisite scallops and a 1995 Calon Segur wine ("a very good year"), Gaydamak launched into long-winded asides, to explain his wealth and success.

He has recently ventured into sports—a bit like the Russian "oligarch" Roman Abramovich, who bought Chelsea football club in England (where, recently, Gaydamak's son Alexandre purchased Portsmouth Football Club).[9] Gaydamak had—around the same time as Israel's withdrawal from Gaza—bought the Israeli basketball team Hapoel (which has been called a left-wing team), as well as a (right-wing) soccer team, Betar Jerusalem. He also made a donation to an Arab team. "During the withdrawal from Gaza, these events, in the media, covered almost 50 percent of the noise. It was very sensitive, the

Gaza withdrawal; with my move, I masked half the media problem, because everyone spoke about me. . . . I already did peace in Africa. I will now do it in the Middle East." I let that one go, and we moved on.

He gave another example of his business skills: a trading operation he set up in the Russian republic of Bashkortostan, near the border with Kazakhstan where Gaydamak has oil and phosphate interests. He bought up several mothballed Bashkir fertilizer plants, supplied them with his Kazakh phosphate, and distributed it to grain producers in Bashkortostan. The farmers could not buy the inputs, so he set up a bank to give them credit—in fertilizer. He bought storage silos (his first harvest was underway as we spoke) and the farmers repay him in grain at harvest time (when the price is two rubles per kilo: "the farmer, he is obliged to sell to me because I have the storage capacity . . . in November a kilo of grain will be five rubles"). He ships the grain to Moscow, where he produces eggs. (Grain makes up two-thirds of the eggs price, so he makes more profits.) "That is how my bank became number two in the area, immediately. And that is how, in one year, I created a company worth $200 million in Bashkortostan."

"In the so-called market economies, with all the regulations, the taxation, the legislation about working conditions, there is no way to make money. It is only in countries like Russia, during the period of redistribution of wealth—and it is not yet finished—when you can get a result. So that is Russian money. Russian money is clean money, explainable money. How can you make $50 million in France today? How? Explain to me!"

The reordering of wealth in Russia has been a bit like Angola's after 1990: a small elite obtained huge state assets through mysterious political processes that are a bit like what happens when an oil-rich country must decide how to share out oil revenues. Some compare this phase with the era of the robber barons that the United States went through in the nineteenth century. But there is a huge difference: the robber barons kept their money in the United States; Russia's and Angola's elites, by contrast, had at their disposal a wonderful array of tax havens and opaque shell banks, protected by successive layers of secrecy like the nested Matryoshka doll that I had bought by Red Square, to help them drain their countries of their wealth.[10]

Gaydamak had some meetings and we broke off; the next day, he promised, we would talk about his $5 billion Angola-Russia debt deal, which some regard as one of the great recent scandals of Africa's oil zones.

The next morning his black Bentley hummed quietly up to the hotel entrance. He had done his kung fu and was unshaven, and without bodyguards, and he wore a suit but no tie. We went for breakfast at the Hyatt, where I laid my microphones on the table. First, he wanted to exercise one of his peeves about the "French mentality" and about the baroque world of backstabbing French politics.

It was the preparations for the 2002 French presidential election, he said, that appear to have really caused him trouble. At that time the political party led by French president Jacques Chirac was in disarray; it had got just 12 percent of the vote in earlier European elections,[11] defeated not only by the Socialists, but also by a new right-wing party led by the French interior minister Charles Pasqua.[12] Pasqua had split the French right wing, and Chirac needed to stop him at all costs.

Pasqua had helped facilitate Gaydamak's and his partner Pierre Falcone's 1993–1994 Angola arms-for-oil operation in France, and in June 2000, two years before the polls, French police handed over information about the deals to French judges. In December they arrested Falcone and issued a summons for Gaydamak, for which he did not turn up, triggering the international arrest warrant. (Falcone, after staying in prison for a year, was released on the largest bail in French legal history, paid by the Angolan state oil company, Sonangol.)[13] In June and November of 2000, Falcone's Arizona-based wife Sonia made a $100,000 donation to the U.S. Republican Party. (The money was returned after Falcone's detention, "to avoid the appearance of impropriety," the Republican National Committee said.)[14] Angolagate implicated Pasqua, and if this was the plot, as Gaydamak says, it worked: Pasqua's party fell apart that month and his challenge to Chirac collapsed. And this, Gaydamak said, is why he was really targeted. "Pasqua was obliged to withdraw his candidacy from the presidential election in 2002 because he was accused in my Angolagate case. That was the only reason." (This is plausible; French fiscal police obtained the information about Angolagate in December 1996, but curiously sat on it until June 2000.[15])

French presidential elections happen in two rounds; several parties fight the first round, and the top two from that contest then fight the second round. (This gives the winner a stronger mandate, as they automatically get more than 50 percent of the vote.) In the 2002 elections Chirac, president since 1995, faced the Socialist prime minister Lionel Jospin, and both were deeply unpopular. Pasqua had previously split the right-wing vote, helping Jospin. But now, with Pasqua discredited, Chirac pulled ahead in the first round and Jospin came third to the far-right anti-immigrant politician Jean-Marie Le Pen, who was repugnant to many French people. "Pasqua's voters went to Le Pen," Gaydamak said. "In the second round, it was Le Pen and Chirac." France, of course, voted massively for Chirac, if only to keep Le Pen out. Chirac became president.

Now, Gaydamak said, Chirac may still need him because he also faces judicial investigation, from which he is currently protected by presidential immunity. Much may depend on who succeeds him at elections in 2007: Prime Minister Dominque de Villepin, Chirac's ally, or his adversary Nicolas Sarkozy. "What can save Chirac?" Gaydamak asked. "Only if the next president is de Villepin. De Villepin made his career under Chirac's management. De Villepin against Sarkozy. But who is Sarkozy? For many years he was Pasqua's number two. Today, French internal policy—it is just a continuation of Angolagate. And Angolagate is a pure product of French internal politics."

We then turned to the Russian-Angolan debt deal, which chilled Angola's relations with the IMF for years. This, Gaydamak said, with characteristic immodesty, was "probably the world's best rescheduling deal."

It is yet another curious tale from Africa's oil zones, involving money flowing into the cracks in the international financial system, between global borders, out of sight.

In the mid-1990s, about half of Angola's $12 billion foreign debt[16] dated from Soviet-era arms and other sales in the Cold War. "The Soviet Union never . . . thought anything would be repaid," he said. "But the bookkeeping was done." This $6 billion debt was clocking up interest, and was troubling the Russian relationship that the Angolan president still valued highly.

A generic problem faced by many African presidents is that to get a thing done, they often can't easily do it by conventional means. The president cannot just make the national health system run efficiently: if he orders dollars to be pumped into the health ministry, most will simply escape through corrupt leaks. Ask the finance ministry to repay Angola's foreign debts to Russia directly, and you might have the same problem.

For President dos Santos, the oil-price crash of the mid-1980s had provided the germ of a solution. The American oil company Chevron wanted to expand its operations, but Angola's state oil company, Sonangol, could not pay for its share of the investments, and President dos Santos needed all the money he could get to shore up Angola's crumbling Petro-Stalinist edifice. Nobody would lend to this basket case.

So they put together something new: an oil-backed loan.[17] Private banks lent Angola the money, and Angola repaid not through the leaky finance ministry, but instead directly in oil cargoes. This was a mechanism that gave the bankers confidence. Soon, Angola was using these loans not just to finance Sonangol's investments, but also for general government use. As in Congo-Brazzaville, oil-backed borrowing became an addiction: by the mid-1990s Angola was feeding its budget not so much with current oil revenues, but instead with these loans, repaid safely offshore with future oil cargoes.

Sonangol had carved out an area of real efficiency amid the chaos of Angola's economy. (Other oil-rich countries like Nigeria, with more chaotic oil sectors, never won bankers' trust enough to pull off this trick.) As successive Angolan finance ministers flailed around hopelessly, Sonangol was repaying its debts punctually, in full. "Bring in useful, focused high technology or training for the oil industry, and Sonangol will make sure your investment is safe," a foreign businessman in Luanda once told me. "But import second-hand cars—anyone can do that—and you risk being eaten by the sharks."

Oil-backed borrowing became President dos Santos's way to bypass inefficiencies in the economy. He also set up oil-backed credit lines with other countries, which worked like this: Angola provides 20,000 barrels per day of oil to, say, Brazil. Next, a Brazilian construction firm builds a dam in Angola. The Brazilian company need not wait, cap in hand, for corrupt Angolan apparatchiks to authorize payment, but instead is paid directly by the Brazilian government, which deducts the cost from the Angolan oil account. This way, many projects that may not otherwise have been built, were built.

It was one of these secure oil-backed mechanisms that Gaydamak had used in his Angolagate dealings in the early 1990s. And by the time the wheels began turning in 1996 in the great Russian debt deal, Elf was just discovering the first of a series of huge Angolan offshore "elephant" oilfields, turning Angola, for a while, into the world's top exploration hotspot. The bankers, who made huge profits from these oil-backed loans, were scrambling to provide new ones. The loans were being feted as a triumph. Each new such loan created a feeding frenzy among Angolan politicians. Few people then appreciated their dark side.

Gaydamak congratulated himself further on the debt deal. The higher Angola's national debt, he said, the more interest banks charge for new loans. So the Russian debt, though dormant, still cost Angola millions in higher interest charges. It also mattered to Russia, which had taken over the former Soviet Union's debts. For Gaydamak, this was a chance to extend his influence not just in Angola, but in Russia too:

> Russia's debt to other countries was $140 billion.[18] When Russia tries to borrow, other countries take into account the $140 billion. So they lend little, and at a very high interest rate. Russia tried to explain: "But other countries owe us $120 billion. The real debt is 140 minus 120 which is only 20!" But the other countries said, "No. You, Russia, have 140. It is real debt. The other countries will never pay you—it was only for financing ideological penetration. Your debt is 140; the other people (who owe you): it is just theoretical."
>
> So I told the [Russian] finance minister, "Angola will pay you back. You will set a precedent. You will tell them, 'Look! Angola, a poor African country at war: they paid!'" That was Russia's interest in this. And I did it. Angola was the first—and the only. We created some kind of precedent; the other countries took it into consideration. OK, not totally, but there was some consideration. And the finance minister became [the Russian] prime minister. They never forgot that.

The deal, briefly, went as follows. In November 1996, Russia agreed to round Angola's $6 billion debts down to $5 billion, then cut this by 70 percent to $1.5 billion.[19] This was diced into 31 promissory notes issued by Angola and given to Russia, promising $48.3 million every six months from 2001 to 2016, after a five-year grace period. Promissory notes were used, Gaydamak said, because legally this meant the debt has been paid. Now, with a lower debt profile, Angola's interest payments also fell; Gaydamak claims that Angola saved *more* money in interest payments in the next five years than it paid to Russia in the first place. (My calculations do not support his claim,[20] and an IMF official to whom I put Gaydamak's calculation here, and his assertions about Russia's interest in doing the deal, was incredulous, and irritated.)

Next, Gaydamak and his partner, Pierre Falcone, set up a private company, Abalone Investments, with an account at UBS in Geneva. This company, he said, then bought the promissory notes from Russia for $750 million.[21] Angola was to pay Abalone the full $1.5 billion between 1997 and 2004.[22]

This strange deal poisoned Angola's relationship with the IMF for years. It seems that UBS was queasy, too: "Any possible mention of one of the representatives of one or other of the parties," an internal UBS memo said, "in a newspaper article, even if a posteriori this is judged to be unfounded or indeed libelous, would not prevent, in the first instance, a Swiss or particularly a Genevan judge taking an interest in the people mentioned."[23]

Angola started to pay chunks of money into the Abalone account, using oil-backed loans arranged through the huge, secretive oil trader Glencore (they were playing a facilitating role, which was entirely legitimate). From October 1997 to July 2000 Sonangol paid nearly $775 million into the private Abalone account—allowing 16 of the 31 Angolan promissory notes to be redeemed to Angola.[24] Suddenly, with 15 notes left to pay, a Swiss magistrate, responding to information from French magistrates, blocked the 15 notes in February 2001.[25] He had found evidence that Sonangol's money flowed into Abalone's account and then flowed straight out again to an array of private companies, a number of which are unknown to industry insiders: nearly $90 million for Falcone or his company, nearly $50 million to a former Yeltsin oligarch, and $60 million to an account in Gaydamak's name. A fifth of it went to an account called "Treasury Ministry of Finance" in a private bank in Moscow, but the judge doubted that even this belonged to Russia's finance ministry (and the bank collapsed in 1999). The magistrates also thought that other unknown recipients, in Luxembourg and elsewhere, might include top Angolan officials, though much of the information was hidden offshore.

The Angolan government wrote a furious letter to the Swiss president, lambasting the judges for this "hostile act." Angolan state media began a barrage of criticism, warning Angola's critics to "keep quiet," and a government spokesman denounced a "campaign of defamation against Angola by a North American multimillionaire of Hungarian descent," a clear reference to the financier George Soros whose Open Society has supported transparency campaigns focusing on Angola. Russia, too, put pressure on the Swiss to unblock the promissory notes, and in October 2003, the notes were unfrozen (but tens of millions of dollars held by unknown companies paid by Abalone remained blocked).[26] The Swiss investigating magistrate who had pursued Gaydamak was promoted away from the job, and in 2004 a new magistrate closed the case, arguing that neither Russia nor Angola had complained about the deal, and seemed to accept Angola's argument that the money in the private bank accounts of the "dignitaries" were "strategic funds placed abroad in a time of war." Transparency campaigners reacted with outrage.

The twists and turns of the Russian debt deal do not end here. Gaydamak says that his company Abalone's total profit for the Russian-Angolan debt deal was $100 million. In addition to the money that appears to have flowed to him, to Falcone, and to the Angolan "dignitaries," many tens of millions also seem to have flowed into a welter of baffling offshore accounts in Switzerland, the Netherlands, Cyprus, Luxembourg, and Israel.

It did not look good, and I pressed Gaydamak to explain. He gave an intriguing answer. Instead of paying Russia back in cash, he said, he paid in "Russian obligations": miscellaneous Russian debts to others that are traded on secondary debt markets. Buying up these "obligations" and then redeeming them to Russia were a way for him to pay Russia what he owed, and then legally extract extra profits by judging the debt markets well. The $162 million that went into the purportedly Russian treasury account was just part of what he paid Russia, he said. "It was a huge stupidity of those people saying 'we can only see the $162 million,'" he said. "Where is the law that says we should pay Russia from this account only?" Money flowed into so many companies, he said, because buying these "obligations" from just one company all at once would move the debt markets against him; it was better to split the finances up, and act subtly from many directions.

Gaydamak railed against the French investigating magistrate Philippe Courroye who is pursuing him (and leading French investigations into the Iraq oil-for-food scandal). "This Courroye is taking everybody in France by the balls," Gaydamak said, literally coughing out the judge's name. Noting the French word *courroie*, a mechanical belt, he tightened an imaginary belt around his neck and screwed his face up to make the point. "In everyday life, in everyday business, my French problem is always on my back."

He turned to the oil deal in Russia that he had mentioned in passing at the beginning of my interview, which appeared to involve the sale of an oilfield (he would not say which, but said it needed $2 billion just for its desulphurization plant). "Because of my French problem, I cannot work with institutional banks. A human alone like me cannot finance [such an expensive project]. That is why I am obliged to sell [the oilfield]. . . . If it were the property of institutional companies, the price would be at least ten times bigger. I doubled the price of my oilfield that I sold. I did not multiply it by ten—so I lost. Why did I not multiply it by ten? Because of Courroye, Courroye, Courroye."

This brought us closer to what I feel really drives Gaydamak: an ambition—though he will not comment about this—to be one of Israel's (and perhaps Russia's) wealthiest and most influential men. In an article about him entitled "Me? An Oligarch?," an Israeli newspaper quoted him as saying something like that. "Because of the image I have been labeled with, because of the international arrest warrant . . . I feel like I'm running with a heavy load that is holding me back and not allowing me to finish among the first few."[27]

Much of what Gaydamak told me cannot be verified. After checking my recording, I still cannot fathom his explanations for where all that money went. Any root-and-branch investigation would end at the cold stone walls of the tax havens.

Is he a victim of the intestinal politics of the French elites? He surely is, though it stretches credulity to believe that he was an entirely innocent victim. Some things he said clash with other evidence. Gaydamak claims that the 1993–1994 arms-for-oil financing was all aboveboard, yet the French businessman Jean-Bernard Curial, who first injected Angolan President dos Santos's plea for help into the French networks, later testified in court that he pulled out of the deals when he saw that they were "a gigantic fraud . . . a vast cash pump," generating a vast 65 percent margin on the biggest arms contracts. On some products, he said, there was a sevenfold markup.[28]

Angola has punished French commercial interests in Angola, partly in retaliation for the magistrates' probes. "Somebody is acting in bad faith," President dos Santos warned in a speech welcoming the new French ambassador to Luanda in 2001.[29] "Friendship is like a plant; if it is not watered and fertilized, it will dry out." Afterward, a big oil license held by the French oil company Total expired but was not renewed, as would normally be expected. It went instead to a Chinese company, Sinopec, deepening the anxiety—you might say outright fear—of western oil firms operating in Angola, desperate to hang onto their licenses and privileges in the face of fast-rising Asian competition. "They injured the independence of Angola," Gaydamak said of the French investigations. "They injured the dignity of the country."

An important question emerges from all this. Transparency campaigners say that the money from Gaydamak's Russian debt deal allegedly in the private accounts of top Angolan officials is obviously the proceeds of corruption. Angola said that it was "strategic funds" placed there for exceptional reasons at a time of war. Who is right?

Plenty of corruption in Africa's oil zones, like the colossal theft of Nigerian oil billions by the spectacularly venal late Nigerian dictator Sani Abacha, is the product of pure greed. But it is not always entirely so. I am struck by a passage in the book *In the Footsteps of Mr. Kurtz*, by Michela Wrong, about the infamously corrupt former Zairean dictator Mobutu Sese Seko, who reportedly salted away $14 billion in offshore banks. When Swiss and Belgian authorities looked for the fabled stash, they came up with only a few million dollars. The U.S. Treasury found less than $50 million, and even reckoned that Mobutu had serious cash-flow problems. Perhaps the bankers did not want to give up the loot. But Wrong offers a different perspective, a bit like what Omar Bongo said about why millions of dollars flowed through his Citibank accounts in New York:

> No one will ever lay their hands on [Mobutu's] fortune, for the simple reason that it does not exist. A lot of money went through Mobutu's hands. But it went through his hands and didn't stay. It was a clientilist system that gobbled cash. Mobutu's theft was a measure not of greed but of political weakness: he needed the money to remain head of one of Africa's largest, most fractious states. There were bribes to be paid to western businessmen, politicians and journalists, wages for the DSP [presidential armed security], donations to foreign guerrilla groups, gifts to generals, governors and opposition politicians. He needed the money to remain head of one of Africa's most fractious states. It was a very, very expensive business.[30]

Was the money allegedly in top Angolan officials' private offshore accounts from Gaydamak's Russian debt deal a Mobutu-like thing—what the Angolans called "strategic funds"? Or was it more like Abacha's loot: straight, selfish venality, pure and simple?

I do not know which it was. It is probably always a bit of both. But President dos Santos's decision to route Angola's loans through offshore oil-backed routes was at least partly motivated by a desire to get things done by avoiding the treacherous Angolan financial system. Was sending Abalone's money into the offshore accounts of his top officials just another way of doing this, to avoid the corruption at home? If politicians salt money away overseas it can also be because it provides insurance against an uncertain future; if they lose power, they will have the resources to protect themselves and their families from their successors. Is this corruption? By a western definition, it is. But it is also not quite the simple, grasping avarice that is so often caricatured in western media, either.

Looking at corruption this way does not excuse it or make it less harmful; either way, oil money that could have ended up paying Angolan teachers and doctors disappeared offshore. It remains a vicious poison for ordinary Angolans, however you look at it. But if we want to tackle "corruption," we must understand clearly what it is. It is often about greed. But at these levels, it is also about staying in power and getting things done.

The oil-backed system that Gaydamak and others used was a triumph for the bankers and for Angola's rulers. But here is the dark side.

Routing the oil money offshore also meant that the finance ministry had no idea where the money was. Transparency campaigners refer to "Angola's missing billions": the more than $4 billion that the IMF reckoned disappeared from state coffers between 1997 and 2002,[31] sucked into the weird, yawning black hole of the oil-backed loans and the wider oil industry. This money was entirely outside the reach of any kind of democratic controls: it was Angola's rulers' money, to dispose of as they pleased. In 1997, the year Angola paid $300 million into the bank account of Gaydamak's Abalone,[32] the IMF estimated unexplained discrepancies in Angola's budget at a staggering 23 percent of gross domestic product.[33] President dos Santos, Gaydamak, and a widely loathed small circle of people at the Angolan presidency used one secret set of accounts, while the finance ministry and central bank operated another set, subservient and confused. To pay salaries, they had to print money, fostering inflation that even in peacetime in the mid-1990s soared up to 1,000 percent, then to 2,000 percent, and beyond—stealing the value of money from the very pockets of ordinary Angolans. It was a terrible time: Angola was for some time described as "the single worst place on earth for a child to be born"; with more than a quarter of Angolan children born not expected to make it to their fifth birthday.[34]

Angola's oil-backed borrowing was a more sophisticated, better-controlled version of the debt tools that French interests used to control the leaders of André Milongo's Congo-Brazzaville. Angola during its civil war became a wealthier, grander, more sophisticated version of the Congolese "successful failed state," the two-speed nation where the apparatus of government was stripped down to little more than its sovereign, extractive core, leaving the rest to wither. The war was, in the words of the Angolan activist Rafael Marques, "a war of poor people against miserable people," conducted while the rulers lived in fabulous luxury. While the fighting killed many hundreds of thousands in the countryside, the financial instability fostered by the oil-backed loans savagely worsened already desperate living standards everywhere. It was a human catastrophe.

Today Angola's civil war is over. Inflation is falling, the financial sector is cleaner, millions of people displaced by conflict have returned home, and the oil industry is generating torrents of cash—the state budget nearly _doubled_ between 2005 and 2006 alone, to $23 billion, as oil companies ramped up production, as oil prices rose, and as billions of dollars of new oil-backed credits rolled in from China. The economy is growing at over 15 percent a year, and Angola is trumpeting one grand project after another: new international airports, huge industrial irrigation schemes, a grand seafront modernization in the capital—and there has even been talk of a new city near Luanda for up to _four million_ people,[35] built by Chinese companies from scratch. Brazilian, American, British, French, Portuguese, Spanish, Russian, and, above all, Chinese companies are swarming in, rehabilitating railways, building a new refinery, and tipping yet more cash into huge new oil projects. Gaydamak's Terra Verde farm that I visited outside Luanda fits nicely with this oil-fired vision; it is, he boasts, "the world's best agricultural project." He had plans for 12 such farms in all. "I am already the main food producer in Angola. You saw how nice it is." I did see it, and it _was_ nice.

Angola's politicians are offering an oil-fired, technologically advanced sprint to modernity and wealth: sowing the oil money, and borrowing more, to fund the binge of postwar construction. The aim is that the giant projects will create jobs and growth, and build an impressive economic powerhouse in the region to rival South Africa. It is an approach based on master planning, and on the idea that the state, armed with its oil and diamond money, will plan its way out of the mess, dragging Angola's citizens from a tormented past, away from their dirty informal activities, into a bright, clean future. In a strange echo of this, the president has set up his own organization, the Eduardo dos Santos Foundation (FESA, to use its Angolan acronym), into which big oil and diamond companies pay contributions and which promotes visions of scientifically advanced development, as well as jamborees on and around the president's birthday to boost his image. A French academic[36] who studied it described it as "the culmination of the process of privatisation of the Angolan state."

We only need to look back to Fela Kuti's Nigeria of the 1970s oil boom to see that this kind of model has been tried before. It is a focus on the *hardware* of development of the kind that was in vogue in development circles in the 1960s and 1970s that left Africa littered with unproductive white elephants and burdened with debt and accelerating corruption. Development theorists today argue that what matters is less the hardware (although this is of course very important) than the *software* of development: secure contract and property rights, independent judges, a free and vibrant media, and above all, good government. Angola was ranked 103rd out of 104 countries in the World Economic Forum's business competitiveness rankings in 2004.[37] Nearly all the foreign investors outside oil and diamonds are going for construction, banking, import-export, and other activities that depend heavily on tapping government contracts—a kind of parasitic capitalism, if you will. There is nothing wrong with these activities, but they will shrivel if the oil price crashes. There *was* a car assembly project mooted in 2005, but German investigators discovered that this was an international scheme, involving several countries, cooked up by some corrupt Germans to swindle money from Volkswagen.[38] Otherwise, pretty much the only productive industries are like those in Congo-Brazzaville: beer, soft drinks, and cement.

Angola clings to this outdated development model now partly because the country has known little but war since the 1970s, when these theories were in vogue. Angola hasn't had much time to think properly about development since then, and its leaders today are the same people who were in charge in the days of Marxist-Leninist central planning in the 1970s. In some cases they have even dusted off old Bulgarian technical assistance models from the 1970s. The politicians also favor a construction splurge because it provides fabulous opportunities and kickbacks. It comes with a nod and a wink from Angola's foreign partners, too—from Gaydamak to the Brazilian or Chinese governments—who are cashing in on the binge. In the late 1990s, when the oil price crashed, Angola used to listen to the IMF a bit, acting like a grouchy and confused beast responding grumpily to the IMF's prodding for reform. Now oil prices are high and Angola's leaders don't need to listen much any more.

Visit a rural village or walk into one of Luanda's desperate shantytowns, and you will see that the Angolan oil-fired miracle is hollow. The only seven countries ranked worse than Angola in terms of life expectancy are all AIDS-ravaged southern African nations.[39] (Angola has been spared

the worst of AIDS so far because the war checked the free flow of people.)
Economic activity is overwhelmingly focused on Luanda; the government
behaves a bit like a bird in the nest with too many chicks—only the closest
get fed. The oil industry employs perhaps 20,000 Angolans,[40] only one in
seven hundred or so of the population, and its high salaries leach the ministries of their best officials, further eroding the state's capacity. While the
oil-fed army, police, and intelligence services keep restive populations in
check, the elites drink champagne, send their children to private schools in
Lisbon and America, and sail in oceans of secret offshore oil and diamond
wealth, while the president rules with the help of a small number of hugely
wealthy unelected advisors. "They are privatizing our president," an Angolan newspaper lamented. "He is being fought over by half a dozen people
in the presidential inner circle, who are like a pack of famished hyenas."[41]

Much of the rest—from Luanda's sewage-soaked, cholera-ridden
slums, to a countryside riddled with landmines and Marburg fever—might
as well have fallen off the map. As thousands of poor Angolans are swept
out of shanty towns and marketplaces to make way for luxury condomini-
ums and shopping centers, secure property rights—a key to unlocking the
entrepreneurial potential of millions of people—further erode. Like André
Milongo's Congo, Angola is two nations.

Gaydamak's legal problems in France are not over, but he is as active as
ever. He said he still flies to Angola once a month; he is also involved there
in a Chinese railway project; he is helping finance a big hydroelectric dam,
and he retains influence in diamonds. He is surely less dominant now:
competition from Chinese and other players is stronger. There doesn't seem
to be an active link with Pierre Falcone, but I cannot be sure. Outside Angola, he owns the Moscow News; and a French newspaper in 2006 said he
owns one percent of Gazprom (whose market value hit $300 billion in May
2006). To me, Gaydamak admitted to owning three banks, adding, of one
of them: "I did not want to get calls from my bank manager about my
credit card, my overdraft. So I created my own bank. The bank manager—
ha ha ha—I never have to hear his voice blaming me."

He continues to stir up controversies: in one of them, in late 2006, he
vocally opposed plans to hold a Gay Pride march in Jerusalem (less controversially, he also financed tents for refugees during the Israel-Lebanon war
in mid-2006). Yet he had told me he was not politically affiliated in Israel.
"Politicians only do it for themselves . . . left-wing politicians, who say they
care about social problems, or so-called right-wingers—they want, subcon-

ciously, that their wives can tell their school friends that their husband is a minister. I am doing real action. It is not politics."

I challenged him with some of my concerns about Angola's current oil-fired development path. He seemed bored by my question. "Maybe," he offered. How about the IMF's and donors' efforts to get Angola's leaders to clean up and deepen their reforms? I asked. "Why do [the IMF] think they know better?" Gaydamak replied.

And, with a shrug: "I don't know. I don't care."

10

DOKUBO-ASARI

CORRODING THE SOUL OF A NATION

In September 2004, world oil prices pushed above $50 per barrel for the first time. This incremental yet symbolic move was triggered by Alhaji Mujahid Dokubo-Asari, a bearlike Nigerian Muslim militant commanding hundreds of armed men who ply the tropical swamps and creeks of Nigeria's Niger Delta in fast speedboats with mounted machine guns. The delta is a sweltering oil-rich zone on Nigeria's southern coast, about one hundred miles northwest from Equatorial Guinea, and it is riddled with mutually suspicious ethnic groups. Asari's Ijaw tribe is the largest in the delta, and is Nigeria's fourth-largest group—after the Hausa-Fulani, the Yoruba, and the Igbo.

Asari, as he is often known, was threatening Operation Locust Feast: an "all-out war" against the oil infrastructure that pumps more than two million barrels of oil each day into thirsty world markets. "The international oil multinationals and the Nigerian state have invaded our land like locusts," Asari said.[1] "Since they are feasting on us and ravaging our land, we also want to ravage them. All foreign embassies should withdraw their citizens from the Niger Delta . . . anyone who assists the Nigerian state to make money in Ijawland will be seen as a collaborator and an enemy and will be targeted."

Asari's threats were the latest ingredient in a cauldron of anxieties for the world's oil consumers. Skittish energy markets were as taut as an African drum, pulled from one end by surging demand from western and

189

Asian economies, and from the other end by supply shortfalls after years of lackluster investment and disruptions in Iraq. "The balance of world oil supply and demand has become so precarious," U.S. Federal Reserve chairman Alan Greenspan said, "that even small acts of sabotage or local insurrection have a significant impact on prices."[2]

The delta is a zone of major strategic interest to the West: its 35 billion barrels of proven crude oil and gas reserves are bigger than all of the U.S. reserves, and it currently supplies nearly 10 percent of all U.S. crude oil imports.[3] Some call this America's spare fuel tank: the delta's light, sweet oil is ideal for making motor fuels and lies just across the Atlantic from American refineries. Nigeria's once-stranded gas is now hissing into western kitchens and bathrooms in fast-rising quantities as advances in technology and cheaper transport allow it to be liquefied and loaded onto ships.

For years, western politicians paid relatively little heed to chaos and poverty here. "Africa," presidential hopeful George W. Bush said in 2000, "doesn't fit into the national strategic interests as far as I can see."[4] Even after September 11, 2001, the paternalistic neglect continued for a while: the delta's inhabitants were not rabid Muslims, so it would not be so hard, policy makers thought, to keep this bountiful tropical spigot open.

Yet by 2002 complacency was giving way to unease. The Institute for Advanced Strategic and Political Studies, an influential Israeli-American think tank, suggested (like the Carter Doctrine of 1980, which endorsed using U.S. military force in the Persian Gulf where necessary) that the United States should establish a military subcommand in this region. "Failure to address the issue," it wrote, "could act as an inadvertent incentive for U.S. rivals such as China, adversaries such as Libya, and terrorist organizations like al-Qaeda to secure political, diplomatic, and economic presence in parts of Africa."[5] That year a U.S. congressman pointed out that seven out of eight of the billion-barrel discoveries in the world in the previous year had been made in West Africa. "In view of Africa's increasing strategic importance to the United States," he said, "the time has come to redefine Africa as a region of Vital Strategic Interest."[6] Western policy makers were waking up. If they wanted the oil and gas to keep flowing, they should understand better why violent radicals like Asari were becoming so prominent.

Alhaji Mujahid Dokubo-Asari was born in June 1964 in oil-rich Rivers State, which local car license plates boast is the "Treasure Base of the Nation." He was first of six children in an upstanding family: his father was a high court judge, and he boasts Ijaw kings, queens, and princesses as ancestors, as well as characters like "Opuwari Alabo Chief Tom Big Harry, Head of Ombo Group of War Canoe Houses of Kalabari Kingdom."[7]

Asari learned Ijaw history and the importance of his ancestors from a strict father and from his grandmother, a rather austere aristocrat to whom his father sent him to live with for eight years. "My grandmother was not a sociable person," Asari remembers. "She was always indoors. She was always warning us that anybody who goes out—if he does not bring trouble, he will bring shame. But I was very rascally and I was the only person who had the courage to question her decisions; every other person was afraid of her. She taught me how to behave, the people one ought to associate with, those who are equal to you and those who are not equal to you." He became a born-again Christian and joined the Deeper Life church, and went to good schools in the oil town of Port Harcourt, the capital of Rivers State. Perhaps some of his grandmother's nature rubbed off. "I never had girlfriends. I never went to parties, and we were not allowed to socialize. In primary and secondary schools there were beautiful girls, but one couldn't talk to them. By my upbringing I wasn't given the opportunity to even try. In fact, I didn't know what to say to girls then."

He attended two universities in Port Harcourt, and in nearby Calabar. Patrick Naagbanton, an ethnic Ogoni journalist who studied with Asari in the late 1980s, remembers him as calm and charismatic, but contrarian. The campus was divided at the time between Trotskyists and Stalinists; although Asari's Ijaw nationalist leanings did not fit these western categories, he became a Trotskyist, perhaps influenced by a trade unionist uncle who had given him Soviet pamphlets as a child. Asari was "a very brilliant boy, reading books we had never seen. A lot of us were angry and we had beards; I had one," Naagbanton chuckled. "We would argue and submit government policy to dialectical analysis. We were not talking about oil, but about building a socialist state. It was very vibrant on campus.

"Then, when communism collapsed in Moscow, a lot of Marxists were unhappy. We are now lamenting the loss of the student movement. Now campus is just a bunch of thieves, criminals, gangsters, and brigands." The two later fell out after Naagbanton supported one of Asari's rivals to lead an Ijaw youth group. "Asari said that if I came out to the creeks, he would shoot me."

In his third year, after clashing with the authorities, Asari left the university and traveled abroad, where, he says, he trained in guerrilla warfare for a year. He will not say exactly where, but some news reports speculate that it was in Libya and Afghanistan. He returned to the Niger Delta and tapped into a simmering fury in what is—paradoxically, given the wealth it produces—Nigeria's poorest region.[8] He also—unusually for the Christian-ized delta—converted to Islam in 1988. Just as Marxism-Leninism had stiffened anticolonial struggles, so Islam included revolutionary elements that Asari liked. "I found Islam the only religion compatible with revolu-tion," he said.[9] "Other religions talk about forbearance, forgiveness, love and peace, endurance in the face of oppression. Islam is the only religion that says that the oppressed must confront the oppressors." He said he has been emir for the Rivers State Muslim Students' Society, and also imam of the Port Harcourt Central Prison.

His Islamic attitudes seem to have melded intriguingly with his conservative upbringing and his Ijaw nationalism. He married in 1990 and took another wife in 1996 (later admitting that he had a "tendency of having four wives," adding that he was more in love with the struggle than with anything else, including his wives). Asari has claimed: "Today in most of our villages you'll find out that most of our ladies have gone into prostitution. Things that were unheard of before like homosexual-ity are being practiced. The Ijaw man is slowly being killed by the Nigerian state."

Among his heroes he cites not only Nelson Mandela but also Osama bin Laden; he has put up at least one poster of the Saudi militant, and even named one of his children Osama.[10] Mandela and Bin Laden "fought against the arrogance of men who were playing God," Asari said. "Apartheid was, 'God gave the white man authority to rule the black man because the black man is not human'; for America it's, 'our civilization is superior . . . so we must impose our way of life, our civilization, on all peo-ple, whether they like it or not.' Oh no no, it is not possible! Keep your way of life and allow other people to keep theirs. That is what Osama is fight-ing. And that's what we are fighting; the Nigerian usurpation of the right of the Ijaw people, the destruction of our environment, our culture. The crux of the matter is oppression."

He denies having any links to al-Qaeda—saying his is a nationalist Ijaw struggle while theirs is religious—yet he admits having some goals in common. "I share some of their aspirations to starve the arrogance of the United States in making the people of the world slaves. The tree of liberty is watered by the blood of the martyrs," he said. "Should they continue to enslave us, we will have no choice but return to the trenches. Then rivers of blood will flow. . . . Dialogue will lead us nowhere."[11]

By the time of Operation Locust Feast in 2004, however, Asari seemed to be leading a contradictory life, austere and flamboyant at the same time. He said that he prayed and listened to Arabic recitations at five A.M., then worked until midnight. "I am a very principled person and I live a near perfect life," he boasted.[12] Yet he was also flaunting great wealth, and cheerfully admitted tapping into oil pipelines. "We are only taking what belongs to us," he told journalists, whom he invited to his lairs in the Niger Delta's creeks and swamps and into his offices in Port Harcourt. He brazenly drove a Hummer jeep, and a luxury Lincoln Navigator with a television in the back.[13] In interviews, Asari would display his bullet wounds, and would prominently place magazines with his picture on the cover; visitors noted that his movie collection ran to Sylvester Stallone and *Mortal Kombat*.

He was clearly proud of his fearsome reputation, too. "One day," he told an interviewer, "police were harassing my wife at a checkpoint. My little daughter—a very small child who looks like a football—opened the car and came out. She went to the policeman and said, "uncle policeman, uncle policeman, I am not afraid of you oh. My father will shoot you!" So the policeman just said, "Madam, madam please go, go with her."

If many Ijaws saw Asari as their Robin Hood, others are less kind. "Asari's influence was huge, but did he use it properly?" asked a western researcher, who had met Asari on several occasions but declined to be identified. "No. I'd call him the playground bully. When Asari—it could have been anyone—said militancy is the answer, people said, yes—a champion! And they followed him."

I first visited Nigeria in 2000, relatively late in my African oil tour. Most of the publications that I wrote for had permanent correspondents in Lagos, so it was harder to find money to travel there.

In the countries that I was more used to—Angola or Gabon, say—the oil is mostly offshore, so its corrosive effects happen in secret arenas inside finance or oil ministries, which are hard to see. But the Niger Delta is different: the oil and gas wells sprout among towns and villages housing up to 20 million people, so the problems generic to oil are in plainer view, and locals instinctively understand far better the dynamics of the trouble that oil can bring. It is here, in the delta's malarial landscapes, that I have probably learned most about how oil corrupts and subverts the very essence of what it means to be a human in society. The end of this last sentence may sound extreme, but you will see from Asari's tale how true it is.

The Niger Delta was formed by the Niger, West Africa's largest river, which flattens out here at Nigeria's southeastern coast into a tangle of creeks and mangrove swamps suffused with fresh and brackish water. Much of the land lies less than two yards above sea level. It takes up less than 8 percent of Nigeria's landmass, but its oil and gas—about half of it produced in operations run by the British/Dutch oil company Shell—generates over 97 percent of Nigeria's exports.[14] It harms the Ijaws, Ogonis, Itsekiris, Urhobos, Igbos, and many other indigenous groups that live here. "Pollution has affected the atmosphere, soil fertility, waterways and mangroves, wildlife, plant life, aqua life, and has resulted in acid rain," the government admits. "Fishing and agriculture are no longer productive enough to feed the area . . . the population is prone to respiratory problems and partial deafness."[15] Oil companies and the government have tried one "community development" project after another, yet the anger just keeps rising. Asari says that the oil companies will "pay for their negligence and genocide."[16]

On this first visit, it seemed safe enough to travel about. I got a local guide and a driver, with the help of a local activist, and we set off into the oil zones.

By a road in a place called Kolo Creek, we stopped next to a wellhead emerging from a concrete plug in the ground. The pipe was quite cold (from depressurization, I suppose—the same mechanism that keeps refrigerators cool) and it hissed softly, dripping condensation onto surrounding thick grass that popped and jumped with grasshoppers. They were perhaps agitated by a thunderstorm we could see nearby, approaching low over the trees. Nearby, 10 or 12 blackened pipes as thick as my leg ran at knee height parallel to the road for a while, then snaked off into the jungle.

We drove on to a flow station, a node in the giant nets of pipelines that oil companies have cast across the Delta since oil was first discovered here in 1956. Here, dissolved gas was separated from the oil, then lit and flared rudely into the atmosphere. A large, tarry-ended pipe pointed a furious tongue of orange flame downward into an earth pit, from where it roared back up into the air to obscure the trees behind. Locals dry tapioca and yams on these flares; from 50 yards away the heat bit into my face. The Delta produces more greenhouse emissions than the rest of sub-Saharan Africa combined, and more gas is flared here than anywhere else in the world.[17] It costs money in the short term to install the infrastructure to stop gas flaring, and this, amid all the scrambling for short-term cash, helps explain why Nigerian politicians, and the oil companies that obey them, are so slow in putting out these polluting flares. In fact, more than a quarter of all gas flaring worldwide happens in Africa. It is yet another way that Africa's corruption-plagued oil and gas industries spread poison into the wider world.

Further along, in a place called Imiringi, an elderly man with white bristles and rheumy eyes shuffled in flip-flops and shabby trousers, offering me a wooden chair under a tin roof, alongside a local Ijaw chief who wore a baseball cap and a tight string of neck beads. They began to unload grievances. "We don't have night here. There is no darkness. The flaring scares the animals away. Our food crops are withering, our plantains and yams are not growing big; the palm kernels are no longer producing. Youngsters go blind at an early age, there are skin problems. We have had no fish in our streams for ten years." The chief launched into a long, byzantine story about a dispute with a nearby village over some land where Shell had found oil, then followed this with a separate tale about recent killings of local youths by the notorious mobile police (who are known to have a tendency to shoot innocent people and then drive nonchalantly away). His tale meandered onward—about jobs that Shell had promised but not delivered, and about a construction contract that Shell had offered to a recently retired Nigerian and to a hated army major who had led a murderous raid on local villages two years earlier.

Next, I met six toughs with muscular necks and holey T-shirts, who hung their legs over chair arms in a sitting room with a shabby sofa, a small television, and a glass case full of books. They gave me schnapps in a dirty glass and told another complex, bitter tale. After some of them had disrupted a Shell flow station, the oil company promised to employ 51 of them as guards and pay cash to each of three nearby communities, plus a bit extra for the chiefs. When the youths went to start work, a discredited local chief turned up with guards and started shooting, injuring three youths. When a local man took them to the hospital, some of the chief's men attacked him with machetes. Police detained the chief for two days then released him; a week after that his men shot dead the youths' good friend, a young local soccer player. The courts did nothing, and now that chief has the most ostentatious house in the village. "We have been in a very sober, sorrowful mood," one said. "We don't know what action to take." They bickered about who should have got what; they wanted more Shell jobs in Imiringi, not in nearby communities; they did not like Igbos and Yorubas getting Shell jobs; they wanted clean water and roads. "The revolution has not ended. The flow stations will be closed down some day.

The Ijaws are out for self-determination and"—all joined in a fist-shaking chorus—"Total Resource Control."

Yet this, too, was a slippery topic. The national budget allocates 13 percent of oil revenues to the state where it is produced, while the "thieves" in federal government get the rest. "This is like saying, 'Give me your shirt and I will give you back a button,'" one of the youths ventured. They were unsure exactly how the 13 percent was calculated, or how best to share it out. "The only thing Shell can do is negotiate with the Ijaws. . . . Let them have an Ijaw Development Commission, headed by an Ijaw." It was the old zero-sum game: how to divide out the more than $20 billion that Nigeria earned that year.[18] The discussion ranged on. At every point, oil was dividing these people from each other.

How is Nigeria's oil money shared out? First, it is split contractually between Nigeria and the oil companies and their contractors. Left-wingers who focus on the exploitation of poor Africans by rich multinationals fixate on this aspect of the division of spoils. This split does matter, but the real mischief begins once that is settled, when Nigerians must decide how to share the money among themselves. The game often turns violent: remember the academic studies that have found that countries without mineral exports have less than one chance in a hundred of civil conflict, while countries where minerals make up a quarter of the gross domestic product (GDP) face a horrifying 20 to 25 percent risk.[19] Nigeria's oil makes up about half of its GDP.

Asari's threats in 2004 did not, of course, emerge from nowhere. Anger has long simmered in the Delta, since at least the time when British, Dutch, and others first came here, looking for slaves. As in francophone Africa, today's oil business has curious parallels with the old slave trade. Two of the most important slave terminals, Bonny and Brass, are now two of Africa's largest oil and gas export points. Traders and raiders caught slaves in the interior, then brought them to local middlemen on the coast, who sold them on to the international slavers. This international trade, just like today's oil trade, fostered intercommunity conflict and attracted floods of guns into the area. As with today's oil trade, western capitalists and a few Nigerians got rich from slavery, while millions of others in the Delta got no benefits, and plenty of the harm.

When slave trading was banned in 1807, British companies turned their attention to palm oil, which was used for making soap, candles, and glycerin for explosives, and which oiled the industrialists' machines (that supplanted the slaves). Locals resented palm oil, too, citing a popular prayer: "May this evil of palm oil not get to our children." In 1895, the Ijaw king Koko—invoking the same Ijaw war god Egbesu whom Asari's fighters would rally to over a century later—led 1,000 fighters in a raid on the Royal Niger Company headquarters at Akassa. They killed 75 people, but Koko was soon defeated and the British tightened their grip. Monopolies or near-monopolies for British companies persisted into the twentieth century, when Shell and BP got an exclusive oil exploration license in 1937.

They struck oil in 1956 in a place called Oloibiri, which is also the birthplace of Isaac Boro, among Asari's greatest Ijaw heroes, who—also invoking the war-like Ijaw god Egbesu—attacked and briefly captured the oil town of Yenagoa in 1966. Boro appointed himself head of state in an independent Niger Delta People's Republic; he declared the oil contracts void, ordered the companies to negotiate only with him, and said that the area had been "purged of non-natives," meaning non-Ijaw Nigerians. The Nigerian army, using boats provided by Shell, recaptured Yenagoa and arrested Boro; he was pardoned and died in 1968 while fighting for federal forces against Igbo-led secessionists in the Biafra war. It was one of Boro's books that had fired Asari up as a youth. "Since then," Asari said, "I have never seen myself as a Nigerian."

Pollution and poverty provoked more unrest, and, by one account, angry communities disrupted oil production more than two hundred times from 1976 to 1988.[20] These incidents were generally small and easily suppressed. But in the late 1980s, collapsing oil prices and painful IMF-inspired austerity—the hangover from the 1970s oil boom—emboldened civil society all across Nigeria.

The pot of grievance was first brought to the boil not by Asari's Ijaws but by a smaller group: half a million Ogonis, led by the firebrand politician and writer Ken Saro-Wiwa. When the Ogonis launched a bill of rights in 1990,[21] rebuking the oil companies for polluting their land and sowing divisions, the plight of the Delta began to find sympathetic ears in the West, where the public had been sensitized to the horror of oil's environmental impact by the Exxon Valdez oil spill disaster of 1989. Saro-Wiwa, like many Africans, understood western sensibilities far better than westerners understood Nigeria, and he exploited the environmentally sensitive mood. His book On a Darkling Plain, about the treatment of minority ethnic groups, added locally to the flammable mix. "The system of revenue allocation and the insensitivity of the Nigerian elite," he wrote, "have turned the [Niger] Delta and its environs into an ecological disaster and dehumanized its inhabitants."[22] Violence spread in Ogoniland, in the first stirrings of a new threat to the world's energy supplies.

In 1990, a local Shell employee learned of an impending protest and called the state police commissioner for urgent "protection." The police responded, and shot 80 Ogoni villagers dead.[23] (A subsequent commission of inquiry absolved Shell of responsibility for the massacre, and blamed the police.) Over the next couple of years, the anger grew in Ogoniland, and in January 1993 an extraordinary 300,000 people—nearly two-thirds of all Ogonis—attended "Ogoni Day" rallies where Saro-Wiwa declared Shell persona non grata and urged minorities to "rise up now and fight for your rights." When Oklahoma-based Wilbros, a Shell contractor, began bulldozing crops for a pipeline, they were met by protestors. Willbros called the police, who shot a man dead and wounded 11 more. Shell offered the village about $17,000 in lump-sum compensation,[24] plus possible individual payments in return for permission from the Ogonis to finish the pipeline. Once again, oil money divided the Ogonis: an Ogoni lawyer remembers that when they went to negotiate with Shell, the people "almost mobbed us. . . . They thought we had sold out and called us vultures. Even my elder brother said, 'Tell me, did Shell give you some money?'"[25]

The protests continued, and Shell had to close its Ogoni operations in 1993, stopping 30,000 barrels per day.[26]

Saro-Wiwa was no saint, and his critics say that his followers used ad-hoc kangaroo courts and vigilantes against rival Ogoni factions. Still, the government also stirred up the divisions and paid Ogoni chiefs to oppose him.

"Shell operations [are] still impossible unless ruthless military operations are carried out," said a memo from a Nigerian official who led the infamous Internal Security Task force in Ogoniland. He called for "wasting operations"[27] and urged "pressure on oil companies for prompt regular inputs."[28]

In May 1994, a mob beat four of these chiefs to death,[29] and the Internal Security Task Force rampaged across Ogoniland, smashing up villages and killing at least 50 people. The task force leader, Paul Okuntimo, later described his tactics[30]:

> I operated in the night. Nobody knew where I was coming from. What I will do is take some detachments of soldiers, they will just stay at the four corners of the town. . . . I will choose about 20 soldiers and give them . . . grenades, explosives: very hard ones. . . . The machine with 500 rounds

will open up. When four or five like that open up and then we are throwing grenades and they are making "eekpuwaaa!" . . . what do you think the people are going to do? We should drive all these boys, these people into the bush with nothing except their pants and the wrapper they are using that night."[31]

Saro-Wiwa and 15 others were arrested and held without access to their lawyers, and accused of ordering the killings of the chiefs. A letter from Saro-Wiwa to his son Ken Wiwa from jail in 1994 illustrates oil's divisive effects once again. "If I can pull this off, I should have started a revolutionary trend in Africa," Saro-Wiwa wrote. "We have to destroy Berlin," he continued, referring to the 1884–85 conference when the great powers carved up Africa, "and remake Africa according to our traditional lines." It might be possible, he added, to "smash the Nigerian jinx over Ogoni and . . . set up a first-class system as I hope to do, as the area is very rich."[32]

Saro-Wiwa was convicted in November 1995, and—despite entreaties from Nelson Mandela, and others—he and eight other Ogonis were hanged.

After the executions, the oil industry in Ogoniland has remained closed, and the baton of resistance has passed to the more than ten million Ijaws, the Delta's largest group. In 1999, Nigerian soldiers, using helicopters taken from a Chevron compound, attacked two small communities in Delta State, killing several Ijaw villagers and injuring several more. Chevron is reported to have told a committee of survivors that this was a "counterattack" by the soldiers after local youths confronted soldiers at a Chevron drilling rig,[33] and said it could not stop the soldiers using its contractors' equipment like this.[34] Later, soldiers and police at a flow station run by Italy's Agip killed nine unarmed youths with machine guns after a protest about community projects. Turmoil deepened.

Oil companies are routinely blamed for any number of things that go wrong in the Delta, but often it is oil-provoked locals who start the trouble: people sometimes damage oil pipelines, so as to cash in on compensation. "Things have got so bad that people will do the most extraordinary things, such as poisoning their own land to get compensation," one observer said. "It comes from a determination not to be swindled." Bronwen Manby, a veteran analyst of the Niger Delta, says that ultimately, others must shoulder most of the blame. "Oil companies, to be sure—not to mention the government—do many terrible things, but it is the oil money that is the problem. The fundamental responsibility lies with us, and our profligate lifestyles in the West."

When the military dictator Sani Abacha died of poisoning in 1998, his successor steered Nigeria toward democracy and called elections the following year. Unfortunately, democracy did not salve the wounds of the people of the Niger Delta, as many had hoped. Instead, the loosening of the military's ironfisted grip created new political spaces, which were rapidly filled by oil money. It entered like a wicked whirlwind.

At first, trouble simmered as unhappily as before. In November 1999, when 12 policemen were killed by an armed gang in Odi Town in oil-rich Bayelsa State, President Obasanjo ordered in the security forces. They killed many hundreds of people—perhaps two thousand, according to Human Rights Watch.[35] Maybe this quieted things down for a time.

But it was ahead of the next election in 2003, once civilian politics were properly embedded, when matters really began to run out of control. (Nigerians bitterly note that the 2003 election date of April 19, or 4/19, echoes section 419 of the Nigerian criminal code, which is famous for denoting Nigerian-style fraud.)

In the run-up to those elections, Peter Odili, the wealthy and flamboyant governor of Rivers State and one of President Obasanjo's strongest supporters, employed Asari to intimidate his opponents. Asari eventually split with Odili, and his men fought bloody battles with Niger Delta Vigilante, another militia led by the feared warlord Ateke Tom. Other vigilantes—such as the Greenlanders, the KKK, the Mafia Lords, and the Vultures—joined either Asari's or Ateke Tom's groups, and Asari's group mostly came out on top. At least 50 militia groups exist in the Delta,[36] and the police admit that they do not have the firepower to beat them. Many began as university fraternities in the early 1990s, then evolved into criminal gangs—or "cults" with traditional cultural underpinnings—that sprang boisterously into the spaces that civilian politics had opened.

Human Rights Watch described one assassination around election time in 2003. "A dark red Mercedes drove up outside the restaurant. A man in a trench coat stepped out, pulled out a gun, and sprayed bullets at the customers. Two women, including a waitress, and two male customers died instantly." One of Asari's men later admitted that they shot eight people in that attack. "I was very close to Odili," Asari later admitted. "So close that I [can] walk into his house at any time."

Ed Stoddard, a Reuters reporter who was in Port Harcort during the polls, remembers vividly: "What I recall was the silence—a sinister atmosphere in any major Nigerian city, where noise and bustle are generally the rule. No queues, no voting, no people on the streets." The Reuters crew was at a polling station, with an armed police escort, when some young toughs showed up in a minivan to intimidate the police guarding the ballots. "The only thing that prevented a confrontation was our presence with the Mobile Police," Stoddard said. "The cops begged us to stay at the polling station. I am sure that they and the election workers abandoned it shortly after we left. It was a shameless rip-off. Michael Moore thinks George Bush stole Florida in 2000. But he ain't seen nothing."[37]

Official data showed that President Obasanjo's ruling party won between 93 and 98 percent of the vote in the delta's four main oil-producing states,[38] on a similarly high turnout. It was a result that Equatorial Guinea's President Obiang would have been proud of. "There is not a single elected party representative at *any* level—local government chairmen, councilors, state house of assembly who isn't from [President Obasanjo's ruling party]," said Chris Newsom, a nongovernmental activist who lives in Port Harcourt.[39] "Rivers is basically a one-party state. There has been a complete collapse of democratic function. People's confidence in the system is going backward, fast. People feel completely alienated, and this drives sympathy for the militants: this is one of the only ways they feel they can get leverage against the government." Militancy has arguably become the core of what you might call modern Nigerian civil society: an organized response by poor citizens to the depredations of their rulers. Perhaps this is like oil-fired militant Islam, only without the Islam.

A central part of the corruption of national politics is "bunkering," or the theft of oil from pipelines. Asari has admitted to doing this, saying that it is fairly easy. "You will just open the pipe, bring the barge, and pump the crude inside." His men sold petroleum condensate tapped from pipelines, which is known locally as "Asari fuel." A consultants' report for Shell in 2003 estimated that a staggering 275,000 to 685,000 barrels *per day* of crude were being stolen by oil thieves in the Delta.[40] At $50 a barrel, this would mean over $5 billion a year—much more than the Angolan rebel leader Jonas Savimbi earned from his diamonds—and used to buy the missiles, armor, and artillery that kept the oil-rich Angolan government and its MiG fighter-bombers at bay for years. Bunkering fuels huge arms flows, amid fierce turf battles for access to this trade. It is no surprise that the U.S. Defense Department is now taking a serious interest in the Niger Delta.

The militants' media profile has created an impression overseas that it is they who are stealing the oil. This is a grand illusion: they take just a small part of the loot from this vast international criminal enterprise. The militants are often not involved directly but extract tolls from passing oil barges, just as riverine communities in past centuries used to tax passing slave ships and palm oil vessels. They provide security for the bunkerers too—and, sometimes, for the oil companies. But the real money flows to bigger fish than the militants, whose noisy activities provide a useful cover. Plenty of the money is captured by international criminal cartels that put stolen oil on barges and then transfer it to bigger tankers on the high seas, which then deliver it to refineries up and down the West African coastline and further afield. This oil probably finds its way into cars and trucks in Europe and America.[41]

Yet several sources in the delta say that much, or even most, of the money from bunkering in fact flows to the Nigerian capital Abuja, to finance factions of Nigerian political parties. In October 2003 the Nigerian navy did impound a ship, the *African Pride,* with 11,000 tons of allegedly stolen oil, near an oil terminal.[42] Yet the vessel mysteriously slipped from navy custody, to the high seas. After a media furor, two admirals were court-martialed and dismissed from the Nigerian navy, but they were never sent to jail. Bunkering continued. A general who sent helicopters to attack bunkerers' ships in February 2006 was fired soon afterward. "Where do these tankers come from?" Asari once asked sarcastically. "These big ships and vessels, where do they come from? The Nigerian navy did not see them. The Nigerian air force did not see them. They are so tiny!"[43]

The deep involvement of Nigerian political parties and prominent politicians may be why the bunkerers—whose tankers are rather easier to spot and catch than, say, Angolan rebels' blood diamonds, which *have* been cracked down on—are almost never caught, and perhaps why oil companies have not seriously impeded the problem. Western policy makers may feel that they need not care much about bunkering; they are just thankful that the stolen oil still flows into world markets, albeit via different routes.

The giant, deadly bunkering industry is in fact a violent, chaotic Nigerianized version of the balloon of offshore dirty oil money that characterized the Elf Affair in Gabon. Money flows to political party factions, and out into the world's tax havens; journalists complain that this is a terrain of smoke and mirrors, where nothing is as it seems. A well-informed source in Washington told me that he believed that at least a bit of this money was being kicked back secretly into the U.S. political system.

On top of all this crime, Nigerian political parties also get huge legal flows of oil money from the budget. The Niger Delta's oil states receive an extra 13 percent of the revenues from local oil activities, on top of the statutory federal allocations to all of Nigeria's 36 states. This 13 percent has enormous effects: Asari's Rivers State, with just three million people,[44] gets nearly six times more money—well over a billion dollars in 2006[45]—than does Lagos State, which is Nigeria's most populous and has four times as many people. Yet public facilities in the delta are pitiful. Fifty years after oil was first discovered at Oloibiri in neighboring Bayelsa State in 1956, the governor is complaining that his state has not yet even been connected up to the Nigerian electricity grid.[46]

Part of the leakage from these huge, oiled state budgets is in plain view, in the palatial homes now sprouting in Port Harcourt suburbs. Another part, like the money from bunkered oil, flows offshore. But plenty, locals say, flows back into the Nigerian national political arena. "The money used for party funding, legally or illegally, seems to be flowing from these parts," said Oronto Douglas, a former Ijaw activist and writer (who later became, to the irritation of fellow activists, information commissioner for one of the oil-rich state governors). "Any party that wants to get power at the center has to get power in these three [oil] states."

This is partly why politics in the Delta is so murky and confused: both the state governors and the militants—who are not obvious bedfellows—share the same goal of squeezing more oil money from the center. When militants clamor for greater "resource control"—for the 13 percent allocated to oil states to be raised to 50 percent and more (or "resource takeover," a more extreme version that Asari supports)—it is hard to be sure who they really answer to. Asari's support for the Rivers State governor before the 2003 elections fed these suspicions, though he and the governor subsequently fell out.

In 2005 Asari was arrested for treason, and, though he remained in prison, his fury and self-belief are apparently undimmed. "President Obasanjo is ready to devour everybody," he said in one court outburst.[47] "He thinks he can break my spirit and make me beg. I will be alive to see him in chains. I am a hunter, I will tame him. We are ready to meet him arms for arms, arsenal for arsenal. If all these do not work, we will use biological weapons. Let him not push us."

In 2006, a Nigerian contact tried to get me and another journalist into Asari's prison cell in the capital Abuja. We got to the prison building, but the guard shift had changed and the replacements, whom my contact did not know, were suspicious about two unknown white men trying to sneak into Asari's cell. State security was alerted, and we decided it was too risky to try again.

Not long after Asari's arrest, a previously unknown rebel group blew up a big oil pipeline. Then, in January 2006, three fast speedboats full of heavily armed Ijaw rebels attacked an offshore Shell facility, forcing a Nigerian navy gunboat to retreat and machine-gunning a Shell support vessel. These shadowy new rebels called themselves the Movement for the Emancipation of the Niger Delta (MEND), and demanded, among other things, Asari's release from jail. Over the next few months they carried out daring coordinated raids on oil infrastructure, kidnapped foreign oil workers, and killed scores of armed policemen in bloody battles in the creeks and swamps.

The rebels seemed to be fulfilling a prediction of Asari's—that removing him would solve nothing. "Isaac Boro was more intelligent than I am; Ken Saro-Wiwa was more intelligent than I am," he once said, in a rare moment of humility. "They killed them and the struggle did not die. There are many people waiting in the wings to take over. It is better to deal with the person you know, because the person you don't know will be more vicious and dangerous. The next one will not talk about *kuli-kuli* [cool-cool] temper!"[48]

MEND gave their hostages bottled water, sardines, and cigarettes before releasing them, and has used the media skillfully, e-mailing warnings in colorful English of impending attacks, using mysterious pseudonyms like Jomo Gbomo or Cynthia Whyte (the latter replied courteously enough when I emailed "her" with inquiries for this book). For a while this became a more formidable, united front than Asari's group had ever been, and their memberships overlap. MEND doesn't seem to have links with armed militants in the Middle East, but some of its tactics seem similar. Days after Nigeria's leaders met Chinese president Hu Jintao to seal some oil contracts, MEND detonated car bombs by remote control near areas affected by the deals. "The Chinese government, by investing in stolen crude, places its citizens in our line of fire," it warned.[49] On another occasion, MEND lambasted an oil company that they claimed had ignored a letter from Asari about an oil spill in 1997. "ExxonMobil," it said, "will be given a taste of traditional but strategic militancy."[50]

On a rare journalists' trip to a MEND hideout,[51] wild-looking men with bare chests and AK-47s or rocket-propelled grenades lined the shore in the steamy sun. Amid reeking mud and marijuana smoke, people disembarking had to wash their heads in the sea to "neutralize every evil-minded person stepping onto the island." Then they pulled off their shirts, drank traditional concoctions, and passed under fetish strings before being offered whisky, Gulder beer, and kola nuts. Nobody was snickering: MEND had taken more than half a million barrels per day out of world oil markets. American security officials who watched televised pictures of MEND's balaclava-wearing fighters saw telltale signs—such as the way they hold their trigger fingers—of serious military training. "All pipelines, flow stations, and crude loading platforms will be targeted for destruction," Gbomo wrote just hours before a violent attack in February 2006.[52] "We are not communists," he wrote, "just a bunch of extremely bitter men."

For a while in 2006, MEND's united front seemed to weaken. When militants threatened to attack the Rivers States governor in July 2006, Asari called from prison to tell them to stop. One group said Asari had become "too soft and too tolerant with the Satanic Rulers of Nigeria," adding that some of his former fighters had started to "romance" with Nigerian politicians.[53] They said they would fight on. Yet despite their irritation with Asari, they still called for his release.

Also in 2006, President Obasanjo announced a new civil construction splurge in the Delta. His government, astonishingly, even gave oil exploration licenses to three companies controlled by militant elements.[54] "I call this the hydra approach," said a British researcher, Joseph Hurst-Croft. "You cut off one source of trouble, you get six months' respite, and then two more grow back to take its place." MEND was apparently annoyed too. "Such handouts from the 'hyenas,'" the group said in a rebuff to Obasanjo's move, "will help no one." Not so long afterward, MEND fighters began to flex their muscles again, killing scores of soldiers in new attacks.

Now, every time the government and the oil companies dole out compensation, lucrative construction contracts, and even oil licenses, the old zero-sum problem rears up again, further fraying the links that hold the Niger Delta societies together. "Shell has become an integral part of the Niger Delta conflict," the Shell consultants' report said in December 2003, noting that the violence was already on a par with conflicts in Colombia or Chechnya.[55] "There is ample evidence to suggest that providing more money to communities may even exacerbate conflict." Chevron has publicly acknowledged that its community aid has been "inadequate, expensive, and divisive."[56]

For those angling for the oil companies' booty, they must stake their claim to being "true indigenes" of the area to qualify for the payments. But who belongs to, and who are the leaders of, the "host communities" where oil is produced? With this question—which dictates who gets the money and who loses out—oil electrifies the issue of who belongs and who doesn't. Chiefs, or aspiring chiefs, pay armed youths to back their political claims.[57] Bloody battles in the oil town of Warri in 1997 also centered on who the true indigenes are. "If the Itsekiris own this place, we become slaves on our own land," an Urhobo chief in Warri once told Human Rights Watch. "People feel marginalized, so you go and pick up a gun just to show that you are alive."[58] Asari's fight with his rival Ateke Tom was another such fight for supremacy.

This is all about belonging, or citizenship. As Ijaws, Ogonis, Itsekiris, and others trumpet their identities and claims to land, the very idea of Nigerian citizenship shrivels, leaving—like the cracks that appear in a lake bed when the water recedes—a fractured landscape. Oil money, by asking people who they are, drives forward the dysfunction, and the threat to world oil supplies. The constant fighting for access to the cash loosens the social glue that binds Nigerians together. Nigeria's very political boundaries have fractured. Three regions at independence in 1960 soon became 4, then 12 states were created in 1967, then 19, 21, 31, and now 36 states, as Nigeria's politicians applied the divide-and-rule trick that Britain perfected in colonial India.[59]

Yet this is just the most visible, top layer of the disintegration. Self-serving politicians support laws that say that only "true indigenes" can get government jobs or contracts, and outsiders who complain are told to "go back to where you came from"—a refrain that immigrants in the rich world might recognize. Over time, as people have moved, and as Nigeria's states have fragmented, more and more Nigerians—especially the most enterprising ones, those who travel and trade and settle elsewhere—become strangers in their own land.

Local governments compete for the cash; oil's divisive character gets to work again at community level, then worms its way on down until it pits village against village, family against family, and even brother against sister. Every day, Nigerian politicians spew oil-based ethnic and regional poison into the newspapers and airwaves. The next president should be from the east! The finance minister should be a northerner! The next governor must be an Ijaw! The contractor who builds this bridge must come from my village! They are all, ultimately, bickering about how to divide up the oil-baked cake. "We are not citizens any more," lamented an Ogoni journalist over a beer in a Port Harcourt hotel. "We are heading for anarchy, statelessness." There can be few better ways to aggravate old hatreds in a political system than to pour oil money in at the top. *This* is how oil subverts what it means to be a human being in society, and how it corrupts Nigerians' shared humanity.

The calls to tribalism, secession, and violence resonate loudly now across the Niger Delta, day after day. "We announce the mandate from the gods of Ijaw land to commence the next phase of strategic resistance against the Nigerian state and her imperialist collaborators," a recent MEND missive said. "So long as the flag-bearer of the Ijaw and Niger Delta struggle Alhaji Mujaheed Dokubo-Asari [is in jail], we will remain sleepless and continue to devise new ways of unleashing havoc on every thread in the fabric of the Nigerian contraption." In nearby Onitsha, in the former Biafra breakaway republic, Biafran pounds have emerged again in local markets.[60]

This recourse to tribalism nourishes Nigeria's legendary corruption, too, by intensifying the scramble. When the governor of oil-rich Bayelsa State was arrested for money laundering in London in 2005 (he allegedly dressed as a woman to slip British house arrest, returning to Nigeria to be impeached), the militants did not repudiate him for stealing their money, but instead welcomed him as a son of the soil. "When he was removed, they played the ethnic card to support him," said Nkoyo Toyo, a Lagos-based activist. "And to play the ethnic card is all too often to play the corrupt card."

The political decompression with the transition to civilian rule has started a new scramble for the riches in which, as ever, the strongest and slipperiest reached the front. It is what the writer Wole Soyinka calls "pandemic prostitution"[61]—or, as a taxi driver put it, the filth rises to the top.

In 2005 Nuhu Aliyu, a respected Nigerian senator and former assistant inspector general of police, said that many criminals he used to pursue were now his esteemed colleagues in parliament. "The Senate Wednesday engaged in a rather dramatic admission of its stinking image," one newspaper wrote of his comments, which caused uproar.[62] "Nuhu Aliyu expressed disappointment that 'there are many of my colleagues here who I had detained for criminal offenses. Unfortunately I found myself with them in the National Assembly.' Some of the senators reacted by shouting 'aha . . . aha . . . aha,' while others hailed him. Not deterred, [another senator] said, 'anybody who says 'aha' is one of them!'" Nuhu Ribadu, Nigeria's anticorruption czar, subsequently said that two-thirds of Nigeria's 36 state governors could face corruption charges when their criminal immunity expires after they leave office following the 2007 elections.[63] In Nigeria you can get rich by being a crook; then you can buy your way into political office, which can make you richer still. It is a vicious circle.

This vicious circle of oil money and politics creates freakish dilemmas for Nigeria's leaders. To pursue an anticorruption agenda it may be necessary to direct corrupt flows of oil money to political godfathers to buy the political support needed to push the anticorruption program through. President Obasanjo's enemies see him as the mastermind behind Nigeria's modern mayhem, though others say that he is just too weak to tame the whirlwind of oil money that has entered the political spaces that multiparty politics opened up.

The Niger Delta is the accelerator of this vicious merry-go-round, poisoning Nigeria's entire political system. "To understand the politics of the delta, it helps to watch *Goodfellas* or *The Godfather*," said a western researcher who did not want to be identified. "It is like Chicago in the 1920s; you know, 'Hit man becomes local government chairman.' Democracy as a facade for the Mafia."

Ultimately this question of cracked citizenship—who belongs, and who doesn't—is the bedrock of a conflict that is running out of control. The Niger Delta is seeing the same kind of mix of alienation and grievance, fed by oil money and surrounded by vast wealth, that motivated and financed Osama bin Laden's attacks on the United States in 2001.

"One day we are all going to roast in this gas and oil," an activist once told the writer Michael Peel.[64] "Because there are people poised to set fire to this thing, there is an insurgency here, there is a revolution."

11

GLOBAL WITNESS

HOOLIGANS AND ROCK STARS

By the mid- and late-1990s I was writing stories about the curious nexus between oil, international diplomacy, and African politics and society. I was able to sell these easily to specialist oil and Africa publications, but they were rather harder to place in mainstream media outlets, which seemed to prefer straighter tales about, say, blow-by-blow accounts of fighting in Angola, or about BP's latest plans.

Western news agencies, the BBC, and the *Financial Times* report from Nigeria in depth, and the BBC and Reuters usually station a foreign correspondent in Luanda, but the rest of this oil-soaked stretch of coastline is neglected: in more exotic terrains like Gabon or Congo Brazzaville they tend to use underpaid and undertrained local stringers with relatively little experience of Big Oil, who are usually under considerable pressure from their leaders to avoid sensitive oil stories. This meant that coverage is often patchy, and linkages between apparently disparate issues are often not made.

In those years one such matter, not directly related to oil, was that Angola's UNITA rebels were digging billions of dollars of diamonds and selling them to big, reputable companies, then using the proceeds to prosecute brutal wars. This was a rather obvious problem that the world appeared to be ignoring.

In 1998 Global Witness, a London-based campaigning outfit I had not heard of, contacted me. They had recently run a successful campaign explaining how timber revenues were financing a war in Cambodia, and they now wanted to look at UNITA's diamonds. I chatted with them, and urged them to look at oil, too. In December 1998 they published a report, *A Rough Trade*, highlighting the purchase of rebel diamonds by big companies. It contained relatively little new information—it was mostly pieced together from news reports and public documents, in a media-friendly format. Another NGO, Partnership Africa Canada, was doing similar work in Sierra Leone where even more brutal, diamond-funded rebels were hacking off ears, limbs, lips, and tongues.[1]

The new diamond campaigns had explosive effects that derived, above all, from their packaging and their simple message: reputable diamond companies were helping African rebels finance brutal wars. *A Rough Trade* caught nice tailwinds, too. It coincided with the collapse of the Angolan peace process, and western countries had by now largely turned against UNITA after the end of the Cold War, and western companies were now focusing on a huge expansion in oil production in Angola. The United Nations, which had screwed up the peace process, was looking for a way to rehabilitate its image in Angola, and supporting a diamond campaign was a good way to start. As Simon Taylor, a Global Witness director, put it, "Clinton moved partly because of the oil, and the French moved because they thought, 'Shit! The Americans will get the oil.'" The international environment was changing too. The world media no longer saw conflicts through Cold War lenses, so were happier to look more deeply into their roots.

"The diamond wars were the secret of the diamond trade until, quite suddenly, they were not," wrote the diamond commentator Matthew Hart.[2] "It seemed to happen in an instant, as if a curtain had been ripped aside and there was the diamond business, spattered with blood, sorting through the goods. Its accuser was a little-known group called Global Witness."

Diamond industry officials I spoke to were furious. Global Witness was crazy; they were naive; they were left-wing sandal-wearing idiots; they had no right. "Global Witness," said one, "is just a bunch of well-intentioned hooligans." *A Rough Trade* was one of the founding documents for the now-famous international campaign against "blood diamonds," which upended the global diamond industry, and brought governments together in search of solutions.

I have chosen to write about Global Witness partly because I don't want to give anyone an impression that I have been alone in delving into this queer world, and partly because their story presents a chance to appraise current western approaches to tackling problems associated with Africa's oil.

Global Witness started up in 1993, when three environmental activists in London—Simon Taylor, Charmian Gooch, and Patrick Alley—decided that nongovernmental organizations (NGOs) were too neatly compartmentalizing the environmental, human rights, and other issues that they were working on. Taylor recalls once learning that South Africa's apartheid army was training Mozambican rebels in the Kruger wildlife park; the rebels were poaching elephants for ivory, using the proceeds to buy arms. "I thought, What kind of issue is that?" he said. "Environmental? Human rights? What is it?"

The three were personally interested in Cambodia, where the notorious Khmer Rouge was funding military activities from logging. "The environmental movement and human rights people were purists, as in, 'It is bad to cut down trees because of tigers and orangutans,'" Alley explained. "This is perfectly valid, but they resisted bringing in root causes. The nexus between natural resources and conflict—no one was crossing those boundaries."

Kicking their Cambodia campaign off, a fellow environmentalist, sensing their enthusiasm, handed over an initial £100 from his pocket in a pub in London. They then raised a few thousand dollars, found a discarded filing cabinet with a working Amstrad computer inside it, and in January 1995 set off for the Thai-Cambodia border, posing as timber buyers. Using pinhole cameras, hidden microphones, and fake business cards, they worked out that the Khmer Rouge was earning over $10 million a month from logging. They published a report that, pretty much for the first time, explicitly linked environmental destruction to the war. Contrary to what many of their peers were doing, they also sought to avoid any appearance that they had an antibusiness agenda.

Their evidence was potent and the effect was dramatic, helping persuade the United States and others to put pressure on the Thai government to close the border to logging. A year later, half the Khmer Rouge had defected (partly because their money was running out) and their brutal "Killing Fields" leader Pol Pot died in a forest hideout in 1998, the year when Global Witness began to look at Angolan diamonds.

Shortly afterward, Global Witness turned to oil, perhaps partly out of a wish to balance their highly politicized diamond work in Angola. I remember a presentation at a think tank in Pretoria around this time, when I asked a speaker whether she thought oil had any linkages with Angola's civil war. Her reply staggered me. "Business is business, and we are talking here about conflict. You are talking about two different things." Nobody was joining the dots.

They contacted me again; we chatted about Angola's oil and I sent them quite a bit of information for their first oil report, *A Crude Awakening,* which was published in December 1999. It called for "full transparency" in the oil sector, to enable Africans—who were, like the rest of us, unable to find out where the oil money was going—to call their leaders to account. Unlike many of their left-wing peers, Global Witness also came down decisively on the side of the IMF, which had been battling for years to eke out more transparency from Angola's leaders. The report also did not advocate a boycott of Big Oil. Angola was by then perhaps the world's hottest place for oil exploration, and Global Witness knew they would never get the oil companies out. "The regular NGO strategy wasn't going to work," Charmian said. "We went to the U.S. and spoke to the NGOs on oil . . . some saw the oil companies as imperialist bastards, others tried to negotiate with them, then got dragged into endless negotiations that led nowhere." Global Witness also appear to have grasped a key point: that westerners wishing to understand Africa's oil zones do well to discard any left-wing (or right-wing) mental baggage. This meant that serious people, notably in Europe, who might otherwise have dismissed them as left-wing woolly thinkers, began to listen.

Despite their unpoliticized stance, the report was still a savage attack on oil firms and on the Angolan and western governments. This set them apart from the better-known Transparency International, which takes a more analytical, less personalized approach, and from Human Rights Watch, which had already (drawing quite heavily on my own research) broached some of the problems of Angola's oil.[3] *A Crude Awakening* had a new, much more inflammatory, flavor. They put some technical details I had sent them about oil-backed loans next to a photo of a stick-thin Angolan girl; and names of the "untouchable oiligarchy" of top Angolans were pasted next to graphics of handguns, oil barrels, and swag-bags sporting dollar signs; there was a cartoon of the Angolan president and UNITA rebel leader Jonas Savimbi standing in piles of tiny human bodies, flinging them in fistfuls at each other.

"*Crude Awakening* created a firestorm in Angola," remembers Arvind Ganesan of Human Rights Watch. "The government went nuts, drawing more attention to it. They denounced and obfuscated; it was a very Soviet response." An Angolan newspaper carried a blank page where its analysis had been censored. A diplomat told me that the activists were "completely nuts-o" and risked destabilizing Angola. A senior oil official's lips and hands quivered with rage as she told me what she hated about the report. Others sniped at its tabloid style.

After their report, I decided not to work with Global Witness any more but to compete with them, as I wanted to write about them, too. I still believe they are, by and large, on the right track, and I have remained on good terms with them, although some of my ideas and insights diverge from theirs. Since *A Crude Awakening*, Global Witness has been at (or near) the forefront of efforts to promote transparency in oil; they are, as someone from a competing NGO told me, "rock stars of the NGO world."

Those engaged in this struggle so far are still mostly Europeans—here I would include the French magistrates and a couple of campaigning groups in Paris, as well as London-based groups like Global Witness—and the issues are now in the newspapers, a bit. If you take media coverage as a yardstick, America is still fast asleep. The battle for transparency has a very long way to go.

Any expert can tell you why transparency is good, and why corruption is bad. But a more interesting question is this: for decades, it has been obvious that corruption and secrecy are devastating Africa—so why has it taken so long for anyone properly to grasp the transparency issue?

The answers to this suggest that Global Witness did not so much propel change, as catch, and reinforce, its wave.

After African states became independent in the 1960s and 1970s, development experts tried one experiment after another. First, they thought that lending for investment in roads and machines would help Africa take off. It didn't. Then they went for investing in education. That didn't seem to do the trick (the scrambling after the collapse of the metaphorical queue in Fela Kuti's Nigeria highlights, for me, a big part of why education alone isn't enough.) The experts decided that Africa's population was growing too fast, so they tried population control. This was not a success, either.

There was a brief upsurge in interest in corruption and transparency with the emergence in 1977 of the U.S. Foreign Corrupt Practices Act, grounded in U.S. horror at Watergate and arms scandals in the Vietnam War. But Cold War realpolitik, and Europe's failure to follow suit (for many of their companies, foreign bribery even remained tax deductible), ground this effort down.[4] Today, almost nobody is getting prosecuted for foreign corruption anywhere.

After the Mexican debt crisis in 1982, the experts began to think that governments themselves—which had borrowed heavily to finance the experts' experiments but then squandered the money—might matter. The answer was to take things out of government hands: cut taxes and spending, privatize everything, and enact IMF-inspired structural adjustment and austerity. Free trade was a big part of this answer—and the oil trade got lumped in with this apparently simple route to prosperity. There was convergence theory too: poor countries grow faster than rich ones, as rich investors pile into undervalued countries and as technology spreads. This was all part of a comforting world view that there will generally be, despite inevitable hiccups, *progress* in the world.

Still, corruption was not really tackled: in the Cold War, secrecy was seen as a legitimate way to keep strategic information from enemies, and forcing governments to disclose information was also seen as taking a stand against them. Corrupt regimes were subsidized and tolerated in the name of ideology, and the problems that emerged were often explained away as being simply the result of the evils of U.S.-Soviet competition. African rulers successfully cast western abhorrence of corruption as a preoccupation of racist outsiders who just do not understand Africa.

What is more, world oil markets were oversupplied in the 1980s, African oil was not considered strategically vital, and the problems afflicting these nations were seen as largely moral and political matters—so westerners often ignored them. By the 1990s, after the Berlin Wall had fallen, victorious pro-marketers argued again that curbing the roles of governments would finally let unfettered market forces work their magic, washing away corruption in the process. But African countries just got poorer.

Another old problem has been that the field of economics has been separated from politics, making it tough to deal with corruption, which is exactly where economics and politics meet. Karin Lissakers, a former U.S. executive to the IMF, said that even in 1994, when she joined the IMF, the word "corruption" did not appear in their texts: the IMF's founding documents said it was an economic, not a political, organization. Corruption is political, so they had no mandate to bother with it.

It should be clear by now how insane this stance was. Who gets rich in the Nigerian free-for-all depends not on the central bank's inflation targets or the optimum budget balance, but on who gets the government contracts. "The IMF is looking at an economic model, when it should be looking at a political model," a former Angolan planning minister once told me. "This has nothing to do with a capitalist system. This is not about production, but about a cake to fight for." In a well-ordered western country, economics and politics are separate. But in Africa's oil zones, economics *is* politics.

Global Witness's Patrick Alley identified another generic reason for the myopia: for years, human rights organizations stuck to human rights, environmentalists stuck to the environment, and so on. "Organizations were hung up on their mandates," he said. "This can be a strength, but it is also inherently inflexible. When the dynamic changes, they tell the funders, 'We do not do x, y, and z.' So they stop."

As the 1990s wore on, however, perceptions began to shift at last. Alassane Ouattara, an Ivory Coast economist who became a deputy IMF managing director, argued powerfully inside the Fund for corruption to be considered, and finally, in 1996, World Bank president James Wolfensohn marked a turning point. "Let us not mince words," he said. "We need to deal with the cancer of corruption."[5] He had placed the issue squarely on the development agenda for the first time at a multilateral institution. Peter Eigen, chairman of Transparency International, calls the World Bank "the most powerful force against corruption in the world nowadays."[6]

The 1997 Asian financial crisis had rammed home the idea that economic liberalization is not enough if it is not accompanied by good governance. Government does not need sweeping away, it needs fixing. The worst disasters in postindependence Angola—civil war, hyperinflation, and corruption—came not from overly strong government, but from government weakness—above all, weak military control and a fragmented economic system with out-of-control spending.

These shifts dovetailed with emerging research by academics such as Paul Collier and Stanford University's Terry Lynn Karl, whose book *Paradox of Plenty* in 1997 helped show why oil windfalls in countries like Venezuela and Nigeria had turned out to be curses, not blessings. By then, the execution in 1995 of the Nigerian Ogoni activist Ken Saro-Wiwa had helped wake the western public up a bit to the turmoil in Africa's oil zones. Other matters, like the rise of information technology and French investigating magistrates' stream of revelations in the Elf Affair, were also filling the transparency campaigners' sails.

It was in probing the several forms of *nexus* between apparently separate worlds that has helped the three campaigners leap from the £100 donation in 1993 to the fast-growing organization of today. Meanwhile, other NGOs, too, are discovering numerous avenues of enquiry at the interstices of our increasingly globalized world. I believe that it is here, in the study of *linkages* between global issues, where NGOs will henceforth carry out their most fruitful work, eclipsing narrow eyes-on-the-furrow approaches that have been a stock in trade for decades.

As Global Witness pushed for transparency in African oil, big challenges appeared. In 2000 the British foreign office called a meeting of big oil companies in Angola, in which they were asked to publish their accounts. Chevron and ExxonMobil did not turn up. This was the year when French magistrates arrested Gaydamak's associate Pierre Falcone, and as stories about Angola's oil, war, corruption, and international politics were starting to emerge in British newspapers.

Then, in February 2001, BP made a move, promising in a letter to publish BP's payments to Angola.[7]

The Angolans reacted with fury. Their riposte came from their state oil company Sonangol, which threatened—in a letter copied to the other oil companies in case they missed the point—to terminate BP's contract in Angola. Sonangol lambasted "organized groups . . . in an orchestrated campaign against Angolan institutions . . . calling for 'pseudo-transparency' of legitimate government actions."[8] For BP, with multi-billion dollars stakes in Angola, this was no small matter. Initially BP seemed to panic, shunning Global Witness—at one meeting Taylor actually noticed a panicked official running away from him. The company has since pulled back on its unilateral commitment.

BP's peers in Angola quietly exulted. ExxonMobil's CEO Lee Raymond said that while he was "not sure" Angolan governance was satisfactory, his company scrupulously observed confidentiality clauses in oil contracts because it was sensitive to local needs; adding that being transparent "could be seen as" influencing how the money is spent—not a proper role for oil companies. He made a sneering contrast with "one company which has run into deep trouble."[9]

Yet the episode still represented progress, of sorts. Global Witness pushed ahead, expanding on the somewhat rough-and-ready, outraged *A Crude Awakening* and beginning a wider dialogue with governments and oil companies.[10]

Sonangol's BP backlash showed that companies would not stick out their necks and publish data alone, so Global Witness shifted its aim away from companies and toward western legislators and regulators. In June 2002 the billionaire investor George Soros[11] joined Global Witness to set up the Publish What You Pay coalition, to ask for oil companies to be forced by western laws or regulation to disclose payments to all governments where they operate. (Currently—and shockingly—companies must only disclose amalgamated data for their global operations; it is impossible to unpick, say, Shell's annual report and work out its payments in Nigeria.) Mandatory requirements to disclose data would level the playing field for companies and would trump confidentiality clauses in the African oil contracts, allowing companies to tell these governments they have no choice but to publish.

Publish What You Pay was a catchy name, but there were other catches, too. Countries like Angola own their oil in the first place: by and large, it is Angola that pays the companies to help produce its oil, and not the other way around.[12] So asking companies to publish what they pay misses most of the cash. "Getting BP to disclose Sonangol's cash flow would be like getting Chevron to disclose Exxon's cash flow," an industry expert said. "It can't be done." To fill in the gaps it would also be necessary to get Sonangol to disclose its payments too. And Sonangol wasn't playing ball. ExxonMobil, TotalFinaElf (now Total), and Chevron also used this and other weaknesses to subtly undermine the campaign.

Yet the broad idea was a powerful one, and the campaign gained traction. Human Rights Watch helped deepen and broaden Global Witness's initial analysis, prompting fresh cries from Angola of "systematic interference in the affairs of a sovereign nation." A confidential IMF document in 2002 that noted unexplained discrepancies in Angola's accounts of over a billion dollars a year—while the United Nations was spending $200 million annually to support a million and more Angolans displaced by the war—helped get more attention.[13] It was impossible to ignore the stark fact that Angola's people had grown dramatically poorer while oil revenues had risen.

While ExxonMobil, Total, and Chevron obfuscated on transparency, BP and Shell, seeing opportunities to distance themselves in the public eye from their competitors, began thinking about disclosing oil data. The British government—impressed by flocks of NGOs joining the Publish What You Pay coalition—publicly praised it, and in 2003 Britain launched the Extractive Industries Transparency Initiative (EITI) aiming not at imposing mandatory requirements on companies to publish their data but at creating incentives to get governments like Angola's to clean up and publish. The IMF came on board for this effort.

Two complementary (and somewhat competitive) approaches were now in place—Publish What You Pay and EITI—each with inherent strengths and weaknesses. The companies love EITI—it takes the pressure off them and puts it onto African governments to disclose—and it is this process that has since made all the running. The NGOs warily backed EITI but argue that its voluntary approach to disclosure—which Taylor says is in part "driven by the Exxon agenda . . . not to join anything that is mandatory"—lets companies off the hook. And, on top of that, those countries where the problem is most serious will be most resistant to change. Under EITI, corrupt governments can choose whether or not to publish data: some have, some haven't, and some have just pretended to.[14] Global Witness promises to snap continuously at EITI's heels to make sure that it stays strong, credible, tight, and relevant, and is not subverted by companies and corrupt rulers. "I am not condemning EITI," Taylor warned. "But I reserve the right to do that. We will pass judgment when it has lived a bit longer."

In short: the NGOs' model, and the complementary EITI model, each have relative strengths and weaknesses, and shared weaknesses too. One shortcoming in both approaches is that they focus on transparency in government oil *revenue*, or the financial flows between the oil industry and national treasuries, and miss where the corruption is often far worse: in government *spending*. Both approaches also miss huge arenas of potential skullduggery lurking in the cost base of the oil industry—the billions of dollars that oil companies can write off against costs.[15] The two approaches are, at best, tackling just part of the problem.

There is another flaw: transparency campaigns rest on the idea that if African citizens know more about where their money is going, they can then "call their rulers to account." And Global Witness' (and others') fiery allegations of corruption have certainly opened up some political spaces in Angola and elsewhere, into which others have followed. However, with a few exceptions,[16] the record here is patchy, as African civil societies are often so weak and fragmented that in many cases it is hard to stir up outrage, except from local groups that are funded by western NGOs that want to drum up interest in the issue.

Global Witness has other detractors. Some individuals and companies have threatened to bring them down in legal disputes (but so far they have not been successfully sued). Charmian Gooch, in an interview in London, said she was particularly irritated with what she called "libel tourism"—where foreign companies use British courts to take legal action because of the ferocity of British libel laws. "There are British lawyers inviting despots to sue in the U.K. That is horrible."

Some of their critics lie closer to home, in the rough-and-tumble world of NGO politics. Some of their peers dislike their cloak-and-dagger research methods; some call them arrogant and uncooperative with others in the field; others complain that they hog much of the funding in the field of transparency, closing down space for alternative approaches to emerge. Some grate at their often rather tabloid presentational style (one person I know calls Global Witness "the rottweiler NGO"), which grabs attention but comes at the expense of proper context and analysis, and some criticize occasional errors that have appeared, notably in their early reports[17] (this goes with the territory: it is always hard to sniff out the *real* story in Africa's secret oil worlds and they have since shaped up). Some allege that in their early attempts to engage on the diamond issue, they have glossed over huge real problems in order to preserve their relationships with all sides.

A certain amount (but not all) of this sniping originates from professional jealousy. (They have found the going tough at times in the United States, where some local NGOs are protecting their turf against the British intruders.)

I was reminded of another yawning weakness in the transparency campaigns in March 2004, a day after Global Witness had published *Time for Transparency*, a document containing colorful allegations about Arcadi Gaydamak, the oil companies, and oil-rich African elites. This was shortly after Chinese President Hu Jintao, in a landmark speech in the Gabonese capital Libreville, had promised a new era in cooperation with Africa: aid "without strings attached." That day, Angola's Angop news service posted two stories at the top of its home page, side by side. One was an attack on a recent western transparency campaign, and the second was a story about a new $2 billion Chinese lending program for Angola to rebuild after the war, on easy terms. "Certain developed countries and international financial institutions condition their support on the constant question of transparency, which is only imposed on the weakest countries. . . . With the Chinese agreement, no degrading conditions were attached," the official news agency reported. Then, in case anyone had missed the point, it added this: "The loan probably marks a turn to the East."

Since then, Chinese companies have flooded into Angola's oil industry and its construction sector, which is now sizzling in a postwar frenzy. China's Sinopec has a joined a huge deepwater partnership with BP, and it took over a big operation from France's Total, which was frozen out of its license because of French magistrates' "Angolagate" investigations. Invigorated by its new Asian friendship and by high oil prices, Angola also began taking on the western oil companies, notably with an attempt (which, until now, has largely failed) to persuade them to route all their payments through Angolan banks.

Yet for all these challenges, small successes are now emerging. Transparency clauses are being inserted into congressional bills on natural resource exploitation, and the IMF and World Bank have taken up the cause with gusto. ExxonMobil, Total, and Chevron are now, timidly, starting to engage. Even Angola, which (at the time of writing) has not joined EITI and still has a tetchy relationship with the IMF, is independently publishing previously secret oil data in detail. This is remarkable, given that Angola's leaders (and several other African leaders) hate the transparency campaigners so much for "neocolonial interference" that I sometimes get the impression they try to do exactly the opposite of whatever it is that Global Witness calls on them to do.

A new chapter is opening—a not insignificant turning point in the world's willingness to tackle the mess in Africa's oil nations. It is not the campaigners' *direct* pressure on Angola (and other oil-rich poor countries) that is pushing them to change. It is the *indirect* pressure—helping change the international political climate, which eventually suffuses back into Africa, via foreign investors and bankers who also see that better transparency and governance in Africa are in their own interests. African countries, in turn, see that being (or at least seeming to be) more transparent is in their own interests too.

We are seeing early stirrings of a part of a global civil society that is beginning to play catch-up with the enormous forces of globalization that financial liberalization has unleashed. Global money flows are almost entirely unregulated, meaning that African rulers can plunder their treasuries—and nobody, anywhere, can stop them. The efforts of groups like Global Witness and Transparency International are preliminary attempts to prod the world's politicians into building up international financial architecture to match, and tame, the enormous forces of financial globalization. Activists on the left who call for the IMF and the World Bank—or the trade that knits countries together—to be abolished are wrong. Activists on the right who call for governments and regulation to be pared back are wrong, too. The IMF and trade rules, and even African governments themselves, must not be weakened or abolished, but instead strengthened and expanded, and reformed to make them more responsive to the needs of the poor. It is essential to fix the disconnects in our increasingly interconnected world.

Global Witness, true to its instincts for finding new linkages, is now seeking to join the dots between disparate campaigns, from Congo to Kazakhstan to Cambodia, to draw common conclusions. "We see the same mistakes again and again," said Alex Yearsley, one of Global Witness's longest-serving officials. "We see different resources, but the same players and the same mechanisms."

At the end of the day, transparency is one thing. Stopping or curbing the looting—the real prize—is another. I will propose some of my own ideas about alternative approaches to tackling this in the final chapter.

CONCLUSION

DRAWING THE POISON

When I first came across corruption in Africa's oil nations, my reaction was a typically British one: visceral, knee-jerk horror, directed principally at African rulers or the oil companies from whose oily teats the rulers suckle. Yet the years following my first mind-bending trip to Equatorial Guinea in 1993 have changed me. This has felt a bit like what Eva Joly describes: "I have had to learn to construct another mental universe, step by step," she wrote.[1] "A parallel economy exists, with its own networks, clients, spheres of influence, and geopolitics." This is the Alice-in-Wonderland world that I glimpsed in Gabon, with tentacles reaching out of African oil deep into politics in large Western democracies. It is the offshore zone where a number of the people and organizations described in this book operate. Dig deeply enough, and you will find that apparently separate matters in African oil often overlap. What is this strange world? It is not the terrain of normally functioning free markets, but something altogether different.

The original disgust I felt at corruption in African oil is blunted by familiarity now, and by a better understanding of these actors' behavior. Maybe this is a bit like studying delinquents: their behavior is often abhorrent, but when you hear about the broken homes that shaped them, your opinion changes. You do not stop trying to tackle the problems, but you approach them differently. My revulsion is now directed less toward these actors and more in two other directions: first, toward oil itself—the dirty, corrosive substance—and, second, toward the *system*—the global financial architecture.

The oil-fueled vicious circles that I saw in Asari's Nigeria are generic: bad leaders corrupt countries, and in corrupt countries bad rulers rise more easily to the top; dependence on oil and gas damages non-mineral industries, making countries ever more dependent on oil and gas; and so on.

223

These are poverty traps. You don't break vicious circles by dabbling. Foreign aid will not do the trick; for serious solutions you need radical surgery. There are many possibilities, but I will focus on just three here.

The first is obvious. We must cut our energy use, drastically and urgently. For the West, this will tackle other problems, too—global warming, energy insecurity, economic instability and risk—and it will shrink this murky offshore world of mischief money that subtly, yet powerfully, hurts us all. One way to cut energy use would be to raise fuel taxes, then, if you like, cut taxes by the same amount elsewhere—on income and food, say. The net result can be that you need not raise net tax rates *at all*. Europeans like me wonder why so many Americans resist this.

To understand the next bit of radical surgery, consider this. It is not unlawful for a United States bank to receive funds derived from alien smuggling, fraud, racketeering, handling stolen property, contraband, environmental crimes, trafficking in women, transport for illegal sexual activity, slave trading—and many other evils.[2]

Can this really be true? I could not believe it at first, and I checked. But it is true. The only catch is this: the crimes must be committed abroad. U.S. anti-money-laundering laws involve a long list of prohibitions on proceeds from crimes—including the above—committed at home, but a very short list for those committed overseas. Welcoming dirty money is profitable for American companies, and it helps fill the current-account deficit. And Europeans, it seems, are hardly better behaved.

Most people are dimly aware of a murky world of offshore finance. But how big is the problem?

Try these for size. The OECD reckons that about *half* of all the world's cross-border trade involves structures for concealing money, involving about 70 tax havens (the French poetically call them "fiscal paradises")[3], as corporations and rich individuals shuffle profits around to avoid taxes and for yet more nefarious reasons.[4] Assets held offshore by rich individuals, beyond the reach of effective taxation, equal one-third of total global assets—or $11 *trillion*, conservatively estimated, costing governments over $250 billion a year in tax revenues. This is more than twice the global aid budget for developing countries.[5] The Cayman Islands (population 45,000) claims to be the world's fifth largest financial center.[6] A U.S. Senate report estimated in 1999—when the problem was smaller—that up to a trillion dollars is laundered through banks each year, half of it through U.S. banks.[7] And so on.

Jersey, Switzerland, the Cayman islands: "Dracula" zones that can't stand the light of exposure—the problem is here, many people believe. But no: New York and London—the playgrounds of Russian oligarchs, Saudi princes, and African dictators—they are tax havens, too! Some now call the City of London the "Laundry of Choice,"[8] with arguably worse controls even than Switzerland. The problem is the barriers erected between countries, which, like semipermeable membranes, let crooks through, but not the police or tax authorities who are chasing them. "The magistrates are like sheriffs in the spaghetti westerns who watch the bandits celebrate on the other side of the Rio Grande," wrote Eva Joly, furious about how tax havens stonewalled her probes. "They taunt us—and there is nothing we can do."[9]

There are basically three forms of dirty money. One is criminal money: from drug dealing, say, or slave trading or terrorism. The next is corrupt money, like the late former Nigerian dictator Sani Abacha's looted oil billions. The third form, commercial money—what our finest companies and richest individuals hide from our tax collectors—is bigger. The point—and this is crucial—is that these three forms of dirty money use *exactly the same* mechanisms and subterfuges: tax havens, shell banks, shielded trusts, anonymous foundations, dummy corporations, mispricing schemes, and the like, all administered by a "pinstripe infrastructure" of mainstream banks, lawyers, and accountants. "Laundered proceeds of drug trafficking, racketeering, corruption, and terrorism tag along with other forms of dirty money to which the United States and Europe lend a welcoming hand," says Raymond Baker, a guest scholar at the Brookings Institution who testified before the U.S. Senate during its investigation into Omar Bongo's millions at Citibank. "These are two rails on the same tracks through the international financial system."[10] U.S. Treasury officials told Baker that in a good year they caught 0.1 percent of the illicit inflows into the country—a 99.9 percent failure rate. A Swiss central banker told him his country's record was probably worse: 99.99 percent. To try to tackle terrorists' financing or the looting of African treasuries without tackling tax evasion by western companies and individuals is to set yourself up to fail.

How can this happen today? It took me a while to work out why it took so long for groups like Transparency International or Global Witness to appear, waking people up to corruption and the importance of transparency. But this one—dirty money—floors me. To explain it to myself I turn to the futuristic comedy trilogy, the *Hitchhiker's Guide to the Galaxy*, which describes an invisibility field known as SEP, the "Somebody Else's Problem" field, which enabled a large spaceship to land in the middle of a busy sports arena without anyone in the crowd noticing. Astonishingly, neither the World Bank nor the IMF has published serious studies quantifying tax evasion.

Françafrique—the spooky, poisonous meld between African oil and French politics that I stumbled across in Gabon—helped corrode French politics and society, and a recent survey found that 85 percent of French people think their country is on the "wrong track."[11] The system was never only about African oil; it was also about French culture, identity, history, business, and many other things. But the oily African slush funds, linked via tax havens into the political systems of France, Germany, and many other countries around the globe lent the system special potency. Before Eva Joly, French people knew something was amiss but didn't have the details, since the media, and the judges, were part of the system. Many people in America today feel similarly uneasy about Saudi money in Washington, and about the grip of big business on the media. I am reminded of the film *Ghostbusters*, where the good folk of New York are disturbed by weird, spooky happenings—until Bill Murray and his ghost-fighter colleagues discover the real source of the mischief: a secret river of supernatural ectoplasm oozing under their city. The ectoplasm is out there, for real: a concealed universe of dirty money, accountable to no country's citizens, where exotic, colorful creatures like Arcadi—and others I am too timid to mention here—thrive. The oily, spooky slime can profoundly harm us; yet its effects are so strange, complex, nebulous, and distant that it is sometimes hard to believe that it even exists.

John Christensen, a former economic advisor to the British tax haven of Jersey who now leads the Tax Justice Network, says that British and American lawyers, accountants, and banks, notably Citibank, HSBC, Standard Chartered, Suez, and the accounting firm KPMG, are the cheerleaders of this giant offshore economy (and they would argue, correctly, that they are acting entirely within the law.) They are joined by U.S. neoconservatives who say that international cooperation on taxation equals signing up to a high-tax cartel, and that avoiding tax is a form of righteous protest. Christensen's agenda, said Grover Norquist, a leading U.S. right-winger "is a direct threat to America's economic interests. The United States is a tax haven, and this policy has helped attract trillions of dollars of job-creating capital to America's economy."[12]

These cries are reminiscent of a statement from the eighteenth century: "From this trade proceeds benefits far outweighing all mischiefs and inconveniences."[13] The quotation is from a slave trader.

In this parallel secret universe the world's biggest and richest individuals and firms—News Corporation, Citigroup, and, yes, ExxonMobil—can quite legally cut themselves loose from pesky full taxation and grow explosively, leaving smaller competitors, who pay their full dues along with the rest of us, choking in their dust. This undermines the very notion of capitalism: the big companies' advantage has nothing to do with the quality or price of what they produce. If you are worried about the power of big global corporations, don't always attack them directly, but attack bank secrecy instead. *This* is the clever way to take on the big fish, using a net that would also snag the Sani Abachas, the Mobutus, the North Koreas, the terrorists, and the drug lords.

Are we talking about just a few bad apples, as the media—terrorized by libel lawyers that dictators and crooks instruct in London—often portray the rare cases that are successfully prosecuted? Africa's Gulf of Guinea will produce 20 or so billion barrels of oil in the next decade, worth perhaps a trillion dollars, of which African governments will take more than half. Could that much money emerging from this swamp of corruption cause mischief at the heart of western democracies, just as Gabon's cash has run amok in France? Half a trillion dollars in the hands of corrupt African rulers, with lobbyists as crowbars for getting into western politics, sounds like another potent threat to me. If half of global trade finance flows through offshore structures, and soon a quarter of America's oil imports will be coming from Africa, I would hazard a guess that we have a systemic and fast-growing problem on our hands.

The dirty world of tax havens is no grand conspiracy, but a decentralized global terrain tucked away in the interstices between states. It is a problem of fragmented global political and economic architecture. National politicians cannot solve this one: only a coordinated international response will do. "Politicians think nationally. Criminals think globally," a French banker once said. "They win, we lose."[14]

Much of the problem could be eliminated with a few well-aimed legislative strokes. One might be to make it a criminal offense for U.S. banks to receive proceeds of overseas slavery, say, or credit fraud. Another might forbid banks from operating in jurisdictions where they are protected from foreign tax and judicial authorities.[15] There are others. What is missing is political will. A global movement is needed now to seek ways to bring it under better democratic control. We should not abolish the World Trade Organization, the United Nations, or the IMF. We should reform them and make them stronger—urgently. More judicial transparency and global cooperation in this would also fundamentally alter political power in poor countries. African politicians loot their treasuries, knowing that once their plunder is offshore, it is safe—theirs for keeps. If they knew that their successors could get the looted cash back, they would suddenly find new incentives for getting the business climate right. *This* is one way to break radically from a vicious circle of oil dependence.

Supporters of offshore secrecy use several arguments in their defense. First, the system delivers prosperity, so why fix it? Next, the problem is already being tackled, with initiatives from the United Nations, the OECD, the European Union, and others. The Cato Institute in Washington articulates a third: "Information exchange raises serious issues of financial privacy and national sovereignty."[16] Fourth, capital flows to lower-tax countries force governments to cut taxes competitively, keeping them on their toes. Finally, closing tax havens will be hard to achieve in a world of sovereign states.

To the first argument I say this: I am all for the free flow of global money—as long as it's legal![17] Is it delivering prosperity? Ask ordinary people in Africa's oil nations or in the Middle East. In this world of growing terrorist threats, are we really comfortable with such yawning global inequalities?

Second, is the problem being fixed? Once again, ask the citizens of Africa's oil zones. As the task forces and other global initiatives move forward, the tide of dirty money is rising. "No one I have ever talked to thinks dirty money is declining," Raymond Baker says. "Indicators point in the other direction."[18] Christensen reckons that about *30 percent* of Africa's GDP disappeared offshore in the late 1990s.

I have seen Cato's third argument many times. Dictators and oil firms *always* wield "national sovereignty" and "confidentiality" to mask secret financial operations. We don't need a global big brother—just global networks of transparency to give crime fighters and tax authorities freedom to probe misdemeanors. Tax evasion, remember, uses *exactly* the same system that helps African rulers steal their citizens' money and helps drug cartels launder their profits.

The fourth argument—that tax competition between countries is good—is entirely bogus, and rests on an economic fallacy. "The notion of the competitiveness of countries, on the model of the competitiveness of companies, is nonsense," wrote the *Financial Times* commentator Martin Wolf.[19] Think about it this way. If a *company* cannot compete, it goes bankrupt and a better one takes its place; this weeds out bad firms and is a source of capitalism's dynamism. But if a *country* cannot "compete," you get a failed state. "I have never heard a coherent argument for tax competition," said Christensen of the Tax Justice Network. "You might as well talk about environmental competition." High taxes or low taxes? I recoil from the idea that competition from tax havens—which Christensen rightly calls "invitations to crime"—should trump what voters want. There are already perfectly good mechanisms for keeping governments on their toes, such as interest rates and inflation: if a government screws up, its currency will fall and the bond markets (and inflation) will punish it. Another good mechanism is called democracy: bad governments get booted out.

The final argument wielded in favor of tax havens—that it will be hard to tackle the problem—is the strongest, because it is so true. It is also the best reason why small or middling-size efforts to fight offshore secrecy—as the Financial Action Task Forces and the like are—will just not do. Dirty money threatens such powerful interests that it needs protestors on the streets, in hundreds of thousands, even millions. One adversary will be what Eva Joly calls the "media-industrial complex"—news outlets that themselves shuffle huge profits around tax havens to avoid taxes,[20] and might well attack campaigns against tax havens.

It is crucial to separate this matter—dirty money—from the question of whether or not free trade is a good thing. Antiglobalization protestors who mingle the offshore issue with calls to roll back free trade or to "hamper international speculation"[21] are muddying the water. Focus on dirty money alone, and you can unite protestors angry about rampaging global multinationals with larger pools of people like me who believe that global free trade is, for all its faults, a good thing in principle. Following debt relief and transparency, this must be the next great issue for global civil society to grapple with. And debt relief is a tiny minnow compared to this monster.

My friend Sam once asked me lightheartedly to sum up, in a word, Africa's deepest problem. A story helps me illustrate what I think is a good answer.

In 2005, I was invited to run a two-week course in São Tomé e Principe to train local journalists about oil. The students from Principe Island declined to travel to São Tomé island for the course; the best oil prospects lay north of Principe, while São Tomé lay south, so the oil, they reasoned, was theirs. They wanted their own training course, on their island. I did not complain, as Principe is a tropical paradise (Bacardi filmed a rum commercial there), and it has some of the world's best sports fishing. The Principe students were polite and diligent, up to the point where we discussed where the oil lay. It was as if I had electrified their chairs. They were on their feet, gesticulating and arguing, furious that the other islanders might get "their" oil.

It was a tiny episode, easily forgotten. But it illustrates the larger issue: mineral resources divide citizens against one another. Regions where locals want to split away from their countries—the Niger Delta, or Angola's Cabinda enclave—are those parts with the oil. It is the old zero-sum game: more oil money for São Tomé means less for Principe, more for northern Principe means less for southern Principe, more for Public Works means less for Commerce, and so on. Finding oil is like dumping itching powder from helicopters, aggravating existing divisions. The divisions are often, but not always, ethnic or religious. Oil's divisive effects are probably most visible in the Niger Delta, where rich, offshore-roaming elites are being separated from the broad populations, and where locals retreat into ethnic enclaves to stake claims to the oil-financed goodies. As they retreat they lose their shared *citizenship:* the very foundation of politics. I am reminded of a toothless old barefoot gent in Kuito in Angola who, when I asked him if Chevron's offshore Kuito oilfield was relevant to him, simply raised his hand above his head and flattened palm downward, to tell me how far out of his depth my question made him feel. He and Chevron live on two different planets. The world is not flat, as the *New York Times* columnist Thomas Friedman suggests: what seems like a level playing field is instead, as the academic Nancy Birdsall notes, a field of craters containing millions of invisible people like the old man in Kuito.

This fragmentation helps me explain why politicians are corrupt and kill growth, when the take from a growing economy could be greater for all. Fragmentation fosters the so-called "tragedy of the commons," a concept that goes back to Aristotle and that explains Chinua Achebe's scramble after a queue collapses. Politicians snaffle the wealth for themselves before their rivals get it, at the expense of long-term policy. The result is a litany of familiar woes: overgrazing, global warming, corruption, and so on. There is no "national interest" to work with, only competing splinters. For me, looking at this through oil-tinted lenses, fragmentation is not just another of the familiar African demons of corruption and conflict. It is the vital source of nutrients that feeds them. Pour oil in at the top of such a society, challenging the splinters to compete for a slice of the pie, and you have a recipe for disaster. Could this be the bloody heart of the conflict in Iraq, an oil-rich and deeply fractured society? I am no Iraq expert, but I bet that it is. Oil is never the *direct* cause; it is always stealthier than that. It just wakes and energizes the old demons, and conjures up a few new ones too.

To tackle the problem, a way must be found to foster trust among the actors and bring them together again. During my research in the African oil zones, I almost lost hope. Oil companies cannot just undo the harm their product does; nor can rulers simply reverse corruption by fiat or through leading by example.

Yet I believe that there is no need for despair, as oil could provide a second tool of radical surgery. An IMF working paper on Nigeria from July 2003 put it like this: "We propose a solution for addressing the resource curse which involves directly distributing the oil revenues to the public. Even with all the difficulties that will no doubt plague its actual implementation, our proposal will, at the very least, be vastly superior to the status quo."[22] The authors then dive off into theories and formulae, which may help explain why it was not much remarked on in the Nigerian press.

This is a crazy idea, you might think: impossible to implement, and impossible to get politicians to accept. I strongly disagree. It is not impossible to implement. What is missing is political will. The prize is potentially so great that it is worth pushing very hard for this. Direct distribution of revenues is a very, very powerful idea. It works for left-wingers, by redistributing money from rich to poor, and it should please right-wingers, too, by taking money away from governments and handing it to private citizens. Want to spread democracy in an oil-rich country? Take the money from politicians and give it to citizens. Political power follows the money.

Today, the experts recommend three main solutions for the resource curse: be more transparent, diversify away from mineral dependence, and create "savings funds," to keep oil money out of local politics. But these have all shown meager results. First, transparency certainly seems to have helped a bit in Africa, but it hasn't made a decisive difference. Second, it is hard to diversify away from a commodity when that commodity attacks and weakens the other sectors that you are trying to promote. Third, savings funds are, as Chad's experience shows, big, tempting pots of cash that can stir up local politics, too. And if a savings fund, by some miracle, stays intact, it then risks perpetuating the resource curse forever after the oil runs out. What a horrible thought! To cap it all, oil-rich countries' savings funds contribute to huge global financial imbalances, which may accentuate the next economic downturn, when it comes.

If the point is to keep the money out of politicians' hands, then distributing oil revenues directly to citizens is the only other legitimate way to do it. That these revenues are often small relative to populations—maybe a dollar a day per person in Nigeria's case (or 30 dollars a day in Equatorial Guinea's)—does not matter so much, as *this is not the main point*. The big benefits are elsewhere.

In Nigeria, there has historically been a relentless splitting of states, a proliferation of administrative divisions to try and solve the problem of deciding who gets what. It does not seem to have worked. Dividing the loot equally, to everyone, would be the ultimate destination of this process, after which point the arguments stop at last. Removing this money from Nigerian politicians' grasping hands would decisively disrupt the endless squabbling, and gut the vast criminal conspiracies roiling at the heart of the Nigerian state. It would also, crucially, make disenfranchised Nigerians citizens again—for the state would be serving them once more. Personally, I could imagine a radical zero percent tax rate on citizens' share of the oil revenues. To get the money for schools, roads, and police forces, the politicians would have to tax the telephone and internet companies, hairdressers and taxi drivers that spring up to cater for the newly wealthier citizens' needs. It would become urgent for politicians to get the business climate right—meaning better lives for all. Citizens wanting to get ahead would no longer look upward to their rulers, like chicks in a nest looking to their mother, but sideways to their fellow citizens. They would trade and deal with each other more, helping rebuild the trust that oil has eroded.

This matters not only for Africans, but it should concern everyone in the West, too. Think of Hugo Chavez of Venezuela, Mahmoud Ahmadinejad of Iran, Osama bin Laden, Dokubo-Asari—and many other characters from oil-rich states: why do they hate the West so much, especially now that their governments have wrested the main oil profits away from western companies? Part of the answer is sometimes religion; another part is that they can afford to snub their noses at the West, as their oil protects them. But there is another, subtler reason. This book illustrates that there are two kinds of money: normal money in normal capitalist countries, on the one hand, and this mischief money from "rents" like oil or gas, on the other. This is reminiscent of the distinction I made in the Congo chapter between the goose that lays solid gold eggs (good money, to be treated carefully), and the gold eggs themselves (which you can boil, drop, or hit with hammers, without too much adverse effect; potentially bad money). The West has taken the good money and left these countries with the bad kind. No wonder their citizens are angry.

If the oil money was paid to their citizens directly, bad money would become good, and they would hate us less. And if, as the Elf affair shows, poor oil states can seriously threaten our own democracies, then distributing the money to African citizens neutralizes this threat, by deflating these mischief-making balloons of oil cash. As a bonus, citizens would be less likely to disrupt the industry that directly feeds them. This would enhance the West's energy security.

Could this really be put into practice? African politicians would detest it, for it would hand "their" money to ordinary Africans. Even today's African "civil society" activists might oppose it: many of them make a living from the current set-up and might lose out in such a revolution. "IMF Working Paper tosh!" sniffed a well-known member of Nigerian civil society to whom I mentioned the 2003 IMF Working Paper. (He is now moving into politics.) "The politicians would find a way to steal it anyway!" is another objection. Or giving people free money makes them lazy. Or it is unworkable in Africa. Or it would cause inflation.

There are answers to all of these. First, the effort needed to design and implement this would be small compared to the challenge of dealing with the growing global threats that emanate from today's tormented oil states. Second, these countries already *do* this rather well, if in a limited way: by paying millions of civil service salaries, even in remote areas. These salaries, which trickle down, stand between millions of Africans and utter poverty.

To get politicians to accept reform, you would need huge grass-roots political action, with international support. Yet African civil societies are weak, because they are, like the politics, splintered. This issue, for once, overturns the fragmentation: in this fight ordinary Yorubas, Igbos, Ijaws, and Hausas would find that they have a powerful shared, unifying goal for a change, and they would be *very* eager to cooperate, in their millions. Even if direct distribution would not fully eliminate the rationale of giving more money to the residents of oil-rich areas, it would neuter the vast, oily criminal networks that feed off and drive these conflicts. Poor countries routinely achieve countrywide vaccination campaigns, voter registration, and other mass grassroots schemes, and rich places like Alaska already *do* distribute oil income directly to citizens. With modern technology—fingerprint and iris recognition and the like—direct distribution should get easier, from a practical point of view. As for the idea that giving people free money makes them lazy: people in poor countries are idle not because they are lazy, but because they lack opportunities. Fix the business climate, and they will work.

How could the international community back this, if it has so little influence? A good way might be with a demonstration model, perhaps using foreign aid instead of oil money. The clean slate in Iraq after the American invasion would have been an ideal opportunity. Unfortunately, it may be too late now to try. "The Iraqi people should embed in their constitution an arrangement for the direct distribution of oil revenues to all Iraqi households," two academics wrote in a banner article in the journal *Foreign Affairs* in 2004.[23] "Such systems minimize opportunities for corruption and misappropriation, since windfall revenue stays out of the hands of public officials . . . success in Iraq would provide a powerful example for other resource-rich countries to follow . . . change, even radical change, is less risky than maintaining the status quo, in which oil continues to wreak the kind of damage it has so often around the world."

A couple more points to end, if I may.

When Hu Jintao, the Chinese president, stepped off a plane in Libreville in February 2002 and announced a new partnership with Africa "without political strings," the West began to wake up to a new challenge. Soon afterward China announced a $2 billion loan for Angola, the precursor of a fast-growing and multilayered new China-Angola love affair, as Chinese companies snapped up Angolan contracts in construction, oil, and diamonds. China was already being criticized for cozying up to the government of Sudan and for ignoring human rights abuses as it took over ownership of Sudanese oilfields, and campaigners now routinely accuse China of neutering the leverage of the IMF, which has been trying to get Angola's leaders to manage their oil bonanza better. Alex Vines, my boss at Chatham House in London, is right to warn against the think tank "Yellow Peril" characterizations of China's engagement, which is multilayered and brings many benefits. Westerners are hypocritical to accuse the Chinese of neocolonialism. And it is not, principally, Chinese engagement that has neutered the IMF. Angola's defiance is instead a result of high energy prices and growing oil production, which insulate it against outside pressure (and which attracted the Chinese, too). Once again it is crude oil itself, not the actors that cluster around it, that is the heart of the problem.

Here is a way to respond to this particular challenge: find ways for the oil money to be distributed directly to African citizens. Western companies lose out to Asian companies partly because the westerners are more squeamish about corruption and other abuses. Deflate much of the corruption, and much of the Asian companies' advantage would presumably disappear too. In the battle for hearts and minds, this, along with transparency, health, and education, needs to be where our long term bets should be.

We also need a new economics of fractured societies. Accepted economic theories work well enough in western societies; these theories are usually painted on one large canvas. But when the canvas is rent into pieces, as it is in many African states, the theories fail or work strangely. In Africa, economics usually *is* politics; rulers decide who gets what. Perhaps recent advances in behavioral economics, game theory, or evolutionary economics, which focus on the behavior of individuals and groups instead of on the behavior of whole economies, will help. In fact, Africa's oil states should help us understand African poverty better. Oil money is a bit like foreign aid, but without the conditions and technical support. It is a useful control experiment. Aid does good in many instances but can cause harm, often for reasons that are related to how money flows through fractured African financial systems and societies.

ExxonMobil likes to say that there is no resource curse, just a governance curse.[24] This is like saying of a heroin addict with criminal tendencies that there is no drug problem, just a criminal problem. They are wrong: the heart of the matter is not rulers' corruption or companies' misbehavior but oil and gas itself. By hollowing out these junkie nations, intensifying competition and conflict between Africans, slowing economic growth, and silently corrupting western politics, it is aggravating some of Africa's worst problems, and stealthily spreading poison around our globalized world.

NOTES

Introduction

1. "Upstream Performance," *Energy Intelligence,* 2006, page 25.
2. In the latest *Financial Times* 500 ranking, General Electric is top, ExxonMobil is second, Microsoft and Citigroup are third and fourth, and Wal-Mart is seventh. See http://news.ft.com/cms/s/e604c4e6–7bc6–11da-ab8e–0000779 e2340,dwp_uuid=4b323b52-d693–11d9-b0a4–00000e2511c8.html.
3. This is calculated over the full life of the oilfield, although current shares are typically lower than this because, for many oilfields, investment costs are still being paid off.
4. Widely quoted. See, for example, http://www.harpers.org/sb-cheney–30209 32092.html.
5. Energy Information Administration, http://tonto.eia.doe.gov/dnav/pet/pet_ move_impcus_a2_nus_ep00_im0_mbbl_m.htm. The United States imported 721 million barrels, or 2 million barrels per day (bpd), from West Africa in 2005, compared to 556 million barrels (1.52 million bpd) from Saudi Arabia and 84 million barrels (230,000 bpd) from Kuwait. The figures are for crude oil and refined oil products. The 2015 forecast is widely quoted. See, for example, http://www.iasps.org/strategic/africawhitepaper.pdf. Around 33% of West Africa's oil went to Asia in 2005.
6. See, for example, http://www.gasandoil.com/goc/company/cna61437.htm.
7. See, for example, "Africa Growing in Importance to World Energy Market," April 2006, http://energy.ihs.com/NR/rdonlyres/DB4721FE–02FA–4D1 D-BA74–0B8DB1DE7464/0/worldwatch0406.pdf.
8. "Chinese President Addresses Gabonese Parliament," Xinhua news agency, February 2, 2004.
9. "China's Global Hunt for Energy," *Foreign Affairs,* September–October 2005, page 28; "Angola Passes Saudi Arabia to Become China's Top Oil Supplier," *Bloomberg News,* March 29, 2006; and *Revista Energia,* Luanda.
10. "Rebel Coalition Threatens Niger Delta," *UPI,* May 22, 2006.

11. Commentary posted underneath story: "EUA: Pyongyang interessada em urânio de Angola, diz estudo," *Angonotícias,* December 11, 2005.

12. IMF Working Paper by Xavier Sala-i-Martin and Arvind Subramanian, "Addressing the Natural Resource Curse: An Illustration from Nigeria," July 2003, page 4.

13. (1) Omar Bongo, Gabon, 1967; (2) Obiang Nguema, Equatorial Guinea, 1979; (3) Jose Eduardo dos Santos, Angola, 1979; (4) Paul Biya, Cameroon, 1982; (5) Maoouya Ould Taya, Mauritania, 1984.

14. See Ricardo Soares, *Petroleum and Politics in the Gulf of Guinea* (Cambridge University [U.K.], May 2005).

15. Nearly five million barrels per day, multiplied by $50/barrel (2005 estimate), multiplied by 365 (days), gives over $90 billion, though at the time of writing, oil prices are above $70, which would give an annual rate of over $100 billion. Total global aid flows are currently $70 to $80 billion. See http://www.guardian.co.uk/g8/story/0,13365,1570090,00.html and http://www.taxjustice.net/cms/upload/pdf/LRB_-_Hooray_Hen-wees_-_6-OCT–2005.pdf.

16. Terry Lynn Karl, *The Paradox of Plenty* (Berkeley: University of California Press, 1997), pages xv and 4.

17. See, for example, the classic Jeffrey D. Sachs and Andrew M. Warner, "Natural Resource Abundance and Economic Growth," Harvard University, November 1997, or www1.worldbank.org/prem/lessons1990s/chaps/Ctrynote 7_AreNaturalResources.pdf.

18. See, for example, Catholic Relief Services, *Bottom of the Barrel: Africa's Oil Boom and the Poor,* page 18.

19. William Easterly, *The Elusive Quest for Growth: Economists' Adventures and Misadventures in the Tropics* (Cambridge, Mass.: MIT Press, 2001), paperback edition, 2002, pages 257–258.

Chapter 1—Fela Kuti

1. From an interview with Bootsy Collins on http://www.jaybabcock.com/bootsyside.html.

2. The second part of the Bootsy Collins quotation is from Michael Veal's *Fela: The Life and Times of an African Musical Icon* (Philadelphia: Temple University Press, 2000), page 89.

3. See Trevor Schoonmaker, editor, *Fela: From West Africa to West Broadway* (New York: Palgrave Macmillan, 2003), page 112.

4. The *Guardian on Sunday,* Lagos, March 16, 1997. Courtesy of the Nigerian Institute of International Affairs.

5. Ibid.

6. The following is the latest IMF data, from "Nigeria: Selected Issues and Statistical Appendix": $27.25 billion in total exports in 2003, with $26.52 billion of it in oil exports.

7. Author's interview with Asiodu, Lagos, 2006.

8. Karl Maier, *This House Has Fallen: Midnight in Nigeria* (New York: PublicAffairs, 2000).

9. Chinua Achebe, *The Trouble with Nigeria* (Oxford: Heinemann, 1983), page 24.

10. Veal, *Fela*, page 45.

11. Crude oil production was roughly 420,000 barrels per day in 1966. See Jedrzej Georg Frynas, *Oil in Nigeria: Conflict and Litigation between Oil Companies and Village Communities* (Hamburg, Germany: Lit Verlag, 2000), pages 17 and 23–25.

12. Eghosa Osaghae, *Crippled Giant: Nigeria since Independence* (London: Hurst, 1998), page 56.

13. Author's interview, Lagos, March 2006. Asiodu was appointed permanent secretary at the oil ministry in April 1971.

14. From 140,000 barrels per day in 1968; IMF, "Nigeria Report," declassified confidential document, September 1, 1971, page 87. EIA data puts total U.S. imports from the Persian Gulf at 895,000 bpd in 1973. From EIA Monthly Energy Review, August 2005.

15. Energy Information Administration, "World Nominal Oil Price Chronology, 1970–2005," in 2005 dollars.

16. Energy Information Administration, "Major Events and Real World Prices, 1970–2005," in 2005 dollars.

17. IMF, "Nigeria: Request for Stand-by Arrangement," declassified confidential document, January 10, 1989, and Frynas, *Oil in Nigeria,* pages 17 and 24–25.

18. See Keith Panter-Brick, editor, *Soldiers and Oil: The Political Transformation of Nigeria* (London: Cass, 1978), page 153.

19. Philip Asiodu was chairman of the national oil company. This version was confirmed by an official involved in the proceedings who did not want to be identified. Author's interview, Lagos, March 2006.

20. Terisa Turner's "Commercial Capitalism," in Panter-Brick, editor, *Soldiers and Oil,* page 191.

21. http://select.nytimes.com/gst/abstract.html?res=F00F14F93F580C738EDDAE0894DB404482.

22. Author's interview with M. D. Yususu, London, May 2006.

23. IMF, "Nigeria—Recent Economic Developments," September 8, 1981, page 9.

24. Osaghae, *Crippled Giant,* page 50.

25. Ibid., page 97. From 1975 to 1978, the area under active cultivation fell from 18.8 million to 11.0 million hectares.

26. The IMF calculated the fall in total production of millet, yams, maize, cassava, rice, melons, cocoyams, groundnuts, and cotton at 69 percent between the 1975–1976 agricultural season and the 1979–1980 season. IMF, "Nigeria—Recent Economic Developments," September 8, 1981, page 69.

27. The academic was Nziru Nzegwu, in Trevor Schoonmaker, editor, *Fela: From West Africa to West Broadway.*

28. From Trevor Schoonmaker, editor, *Fela: From West Africa to West Broadway.*

29. Veal, *Fela,* page 69.

30. See http://arts.guardian.co.uk/features/story/0,11710,1284327,00.html.

31. Ibid.

32. Achebe, *The Trouble with Nigeria,* page 9.

33. This is widely true, and a result of governments tending to want to "capture the windfalls" by exacting very high marginal tax rates. Publication of contract details is, regrettably, rare, but Shell did publish an example on its Web site for Nigeria, showing how a 200 percent rise in oil prices from $10 a barrel to $30 a barrel translates into only a 113 percent rise in revenues for Shell but a 371 percent rise in revenues for Nigeria.

34. IMF Reports, August 14, 1989, page 2, and May 21, 1985. This is a 72 percent fall in export revenues, while oil prices had fallen only by 60 percent, from $35.60 a barrel in 1980 to $14.20 a barrel in 1986.

35. Interview with Mark Allen, November 17, 2005, Washington.

36. Osaghae, *Crippled Giant,* page 204.

37. Foreign debt grew from 4.5 percent of GDP to 23 percent between 1980 and 1983. IMF Staff Report, June 13, 1986, page 3.

38. IMF, "Nigeria—Recent Economic Developments," August 18, 1989, page 9.

39. See Osaghae, *Crippled Giant,* page 172.

40. IMF Report, May 21, 1985. Total revenue declined by 12.6 percent of GDP during 1980–1983, while total spending fell by only 2 percent of GDP. So the fiscal deficit ballooned from near balance to 11 percent of GDP in 1983.

41. From Table A–9 in Terry Lynn Karl's *The Paradox of Plenty* (Berkeley: University of California Press, 1997).

42. Abdulsalam Abubakar.

43. IMF Working Paper by Xavier Sala-i-Martin and Arvind Subramanian, "Addressing the Natural Resource Curse: An Illustration from Nigeria," July 2003. See page 4.

44. http://ias.berkeley.edu/africa/Events/OilHR/HENDRO-OILFIN-FINAL%20PAPER.pdf.

45. See William Easterly, *The Elusive Quest for Growth,* pages 255–281.

46. See http://news.bbc.co.uk/2/hi/business/4359286.stm.

47. See http://news.bbc.co.uk/2/hi/africa/5129270.stm. In another survey, Obasanjo was the second least popular African leader, after Robert Mugabe of Zimbabwe

48. Carlos Moore, *Fela, Fela: This Bitch of a Life,* translated by Shawna Moore (London: Allison and Busby, 1982).

Chapter 2—Pedro Motú

1. From author's conversations with Bennett; see also interview with Bennett in Douglas Farah, "A Matter of Honor in a Jungle Graveyard," *Washington Post Foreign Service,* May 14, 2001.

2. See, for example, U.S. State Department Equatorial Guinea Human Rights Report, January 31, 1994.

3. From U.N. Human Rights Commission Report E/CN.4/1994/56, which is available at http://www.unhchr.ch/Huridocda/Huridoca.nsf/0/1a4ce1c8a78 d3ea880256732004bcf2f?Opendocument. Artúcio, the UN special rapporteur, was subsequently accused by the government of "violating national sovereignty" and "meddling in national affairs."

4. Max Liniger-Goumaz, *Small Is Not Always Beautiful: The Story of Equatorial Guinea* (London: Rowman & Littlefield, 1988).

5. Amnesty International monthly bulletin, March 1978.

6. See http://www.forestsmonitor.org or Max Liniger-Goumaz, *Guinée équatoriale: 30 ans d'état délinquant nguemiste* (Paris: Harmattan, 1998), page 18.

7. See "Guinée Equatoriale sous la botte d'un clans," *Le monde diplomatique,* July 1994.

8. Liniger-Goumaz, *Small Is Not Always Beautiful,* page 147.

9. See the *Sunday Times,* April 16, 1978, and Karl Maier's interview with Ojukwu in *This House Has Fallen: Midnight in Nigeria* (New York: PublicAffairs, 2000), page 285. See also Max Liniger-Goumaz, *Historical Dictionary of Equatorial Guinea,* second edition (Metuchen, N.J.: Scarecrow, 1988), under "Forsyth." Forsyth has admitted passing money to the plotters, but he said it was only for information. When asked whether he helped plot it, he said his memory was vague: "I don't know whether I thought of it, or someone else." See Adam Roberts, *The Wonga Coup: The British Mercenary Plot to Seize Oil Billions in Africa* (London: Profile Books, 2006), page 35.

10. See, for instance, the report of the U.N. expert Fernando Volio Jiménez, E/CN.4/1992/51, in "La Guinée Equatoriale sous la botte d'un clan," *Le monde diplomatique,* July 1994.

11. See Jeremy Atiyah's article, "192-Part Guide to the World: Equatorial Guinea," *Independent* (U.K.), June 20, 2000, or Liniger-Goumaz, *Guinée équatoriale: 30 ans d'état délinquant nguemiste,* page 47.

12. See "U.S. Oil Firms Entwined in Equatorial Guinea Deals," *Washington Post*, September 7, 2004, available at http://www.washingtonpost.com/wp-dyn/articles/A1101–2004Sep6.html.

13. World Bank, "World Bank Memorandum to the Executive Directors and the President," July 1, 2002. The World Bank said that both the Walter contract and the Block B contract held by Mobil (which obtained it from United Meridian Corporation in 1994) were signed without its knowledge.

14. The World Bank said shortly afterward that all of Equatorial Guinea's oil licenses had been negotiated through closed deals, rather than open tenders. See the internal document compiled by World Bank officials, "Guinea Ecuatorial: El desafío de la riqueza inesperada," September 14, 1998, pages 30 and 35. The internal document also states that acceptable environmental safeguards were not put in place, either.

15. See, for example, "Alexander's Oil and Gas Connections" (http://www.gasandoil.com), February 6, 1998, quoting Mobil spokesman Lloyd Slater: "we have helped them in the past by bringing forward cash payments due to them." Oil officials in Malabo also told me of at least six such renegotiations, not only by Mobil; it is believed that there have been several more.

16. IMF, "IMF Concludes 2001 Article IV Consultation with Equatorial Guinea," October 11, 2001: "Large extra-budgetary expenditures have been financed since 1996 through advances on oil revenue, and the oil companies have been withholding government oil revenue at source to repay these advances."

17. In "Guinea Ecuatorial: El desafío de la riqueza inesperada," page 8; the interest rate is estimated at more than 30 percent in some cases. The IMF estimates the implicit interest rate at 19 percent (IMF Country Report no. 99/113, page 21).

18. IMF Country Report no. 99/113, page 20, notes: "The expressed lack of capacity on the part of the Equatorial Guinean authorities to effectively monitor oil costs properly and to systematically evaluate oil company budget submissions likely makes the emphasis on up-front royalty payments—rather than on tax and profit-sharing payments as in Angola—a convenient arrangement." IMF Country Report no. 99/113 is available online at http://www.imf.org.

19. See, among others, U.S. Department of State, "Country Reports on Human Rights Practices," 1999, available at http://www.state.gov, for a description of activities of APEGESA.

20. Obiang told me this during my interview of him in November 2003, in the presence of the *Jeune Afrique* journalist Jean-Dominique Geslin. In the Omar Bongo Citibank Senate investigation, "U.S. Senate Permanent Subcommittee on Investigations Hearings on Private Banking and Money

Laundering," November 9 and 10, 1999, page 328, there is confirmation of this monopoly.

21. See Senate Hearings—for instance, pages 69 and 328—for an outlining of Obiang's holdings via Abayak. (An ExxonMobil official said Abayak was controlled by the First Lady.)

22. Mobil did not negotiate the original contract; it acquired its interests in Block B from United Meridian Corp. in May 1994.

23. IMF Country Report no. 99/113, pages 63 and 74, estimates government oil revenue at 31bn CFA in 1997, while oil exports were worth 237 bn CFA. While it is common for countries to receive a low share of revenue in the early years, this is a very long way below regional averages.

24. Government revenue shares are usually low in the early years of production, as large initial development costs are paid off, but rarely this low: Angola has been getting an average 25 percent of the value of early oil from its large deep-water projects, after which its share rises dramatically.

25. Wood Mackenzie, "Equatorial Guinea Restructures PSC," *West Africa Upstream Report*, July 1998. The forecast was based on a conservative inflation-adjusted oil price assumption of $16 per barrel; my number has been calculated from a graphical representation of the data.

26. Andrew Latham of Wood Mackenzie, in April 2004, told the author this, and he estimated recoverable reserves in excess of a billion barrels. Also, in "Exxon, Devon boost Equatorial Guinea Zafiro Oil Output 37 Percent," *Reuters*, July 30, 2003, Zafiro's ultimate recoverable reserves stand at "more than a billion barrels." And see http://www.rigzone.com/data/projects/project_detail.asp?project_id=72, which estimates Zafiro's recoverable reserves at 1.3 billion barrels.

27. This figure assumes that the contract's revenue share scales evenly with rising reserves estimates and oil prices; but $600–700 million, multiplied by three (times the reserves), multiplied by three more (times the oil price), comes to about $6 billion. Wood Mackenzie takes inflation into account in its calculation, so a discount rate is not required.

28. From Platts, quoted in EIU Gabon/Equatorial Guinea country report, 2nd quarter 1998, page 28.

29. See IMF Country Report no. 99/113, October 1999, page 20, and "Guinea Ecuatorial: El desafío de la riqueza inesperada," pages 8 and 33. According to the World Bank and IMF, Equatorial Guinea would receive 15–40 percent of the value of its oil from Block B (the Zafiro block), "probably at the lower end of this scale," compared to 45–90 percent for nearby countries.

30. World Bank, "Project Performance Assessment Report Equatorial Guinea— Second Petroleum Technical Assistance Project (Credit 2408-Eg)," July 1, 2002, page 9.

31. Speech at a luncheon in Washington, D.C., for the U.S. Corporate Council of Africa, May 14, 2002.

32. For an excellent dissection of Kapuscinski's work, see John Ryle, "At Play in the Bush of Ghosts," *Times Literary Supplement,* July 27, 2001; the article is available at http://www.richardwebster.net/johnryle.html.

Chapter 3—Abel Abraão

1. See "Mystery of Va.'s First Slaves is Unlocked 400 Years Later," *Washington Post,* September 3, 2006.

2. Renato Aguilar and Åsa Stenman, *Angola 1994: Trying to Break through the Wall* (Gothenburg, Sweden: Department of Economics, Gothenburg University, September 1994). This is one of the only historical documents available detailing official accounts at that time.

3. *Affaire Elf—Affaire d'État; Loïk le Floch-Prigent: Entretiens avec Éric Decouty,* page 56.

4. Angola published its first-ever set of national accounts in 1993, the year I arrived. See Aguilar and Stenman, *Angola 1994: Trying to Break through the Wall.*

5. See "The Dragon of Death Who Had to Be Slain," *Telegraph* (London), February 24, 2002.

6. Steinberg told the author something along these lines in 1998; the quote itself is from http://dir.salon.com/story/ent/feature/2005/08/17/abramoff/index1.html.

7. See http://www.publicintegrity.org/bow/report.aspx?aid=155.

8. Loïk Le Floch-Prigent, *Affaire Elf, affaire d'état,* page 57. This is a slightly condensed version of what he said.

9. See interview with Karen Brutents, Communist Party Central Committee, http://edition.cnn.com/SPECIALS/cold.war/episodes/17/script.html.

10. See Witney W. Schneidman, *Engaging Africa: Washington and the Fall of Portugal's Colonial Empire* (Lanham, Md.: University Press of America, 2004).

11. See, for example, Tony Hodges, *Angola: Anatomy of an Oil State* (Bloomington: Indiana University Press, 2001), page 41.

12. Author's interview with a banker who wished to remain anonymous, November 1999.

13. See the Economist Intelligence Unit (EIU), Country Report, second quarter 1998.

14. Inge Tvedten, *Angola: Struggle for Peace and Reconstruction* (Boulder, Colo.: Westview Press, 1997), page 80. The original account is from a British journalist, David Ottoway.

15. Renato Aguilar and Åsa Stenman, *Angola 1995: Let's Try Again* (Gothenburg, Sweden: Department of Economics, Gothenburg University, June 1995).

16. See http://www.hrw.org/reports/1994/WR94/Africa–01.htm.

17. For an account of Buckingham and Executive Outcomes, see Duncan Campbell, "Marketing the New 'Dogs of War,'" U.S. Center for Public Integrity, October 30, 2002, available at http://www.publicintegrity.org/bow/report.aspx?aid=149.

18. For a firsthand account of the battles at Soyo and the new contract with Façeira, see Jim Hooper, *Bloodsong* (London: HarperCollins, 2003).

19. Global Witness, *All the President's Men,* March 2002, page 13, gives the date of this shipment.

20. See Global Witness, *All the President's Men,* for description of the deal.

21. A modernized version of the AK–47.

22. Pressure from Washington, just ahead of the peace agreement, stopped the army from trying to defeat UNITA fully. This disastrous meddling helped usher in eight more years of trouble and civil war.

23. Author's interview with Abraão, Kuito, September 1993.

Chapter 4—Omar Bongo

1.   A good article about Bongo and France in Africa is the BBC's "France: Superpower or Sugar Daddy?" December 23, 1998, available at http://news.bbc.co.uk/1/hi/special_report/1998/12/98/french_in_africa/235589.stm.

2.   Now officially Omar Bongo Ondimba, after adding "Ondimba" to his name in 2004.

3.   Shell has spent considerable money cleaning up spills from oilfields near its Gamba accumulations; other spills may have happened, too.

4.   IMF Country Report, November 2002.

5.   Author's interview with Jean Ping, 1997.

6.   Bongo won 79 percent of the vote in 2004 and 67 percent in 1998.

7.   See "Lobbyist Sought $9 Million to Set Bush Meeting," November 10, 2005, available at http://www.nytimes.com/2005/11/10/politics/10lobby.html?ex=1289278800&en=564c967b17493798&ei=5088&partner=rss-nyt&emc=rss. See also "Gabon President Demands Retraction, Apology from NY Times," *PRWEB,* January 5, 2006, available at http://www.emediawire.com/releases/2006/1/emw323655.htm. The Bongo-Bush meeting did happen, but there is no evidence Abramoff arranged it.

8.   "Beauty and the Bongo," *National Post* (Canada), February 7, 2004.

9.   Alain Lallemand, "The Field Marshal," *U.S. Center for Public Integrity,* November 15, 2002, available at http://www.publicintegrity.org/bow/report.aspx?aid=155.

10. "Our New Best Friend—Who Needs Saudi Arabia When You've Got São Tomé?" *New Yorker,* October 7, 2002.

11. Omar Bongo, *Blanc comme nègre: Entretiens avec Airy Routier* (Paris: Grasset, 2001), pages 23–26.

12. Bongo, *Blanc comme nègre,* page 114.

13. Ibid., page 39.

14. Martin Meredith, *The State of Africa: A History of Fifty Years of Independence* (London: Free Press, 2005), page 11. The minister was Herbert Morrison, the home secretary in Churchill's wartime coalition.

15. Bongo, *Blanc comme nègre,* pages 37–38.

16. Maurice Robert, *Ministre d'Afrique: Entretiens avec André Renault* (Paris: Seuil, 2004), page 94.

17. Antoine Glaser, Stephen Smith, *Comment la France a perdu l'Afrique* (Paris, Calmann-Lévy, 2005), pages 44–46, 52.

18. Bongo, *Blanc comme nègre,* page 104.

19. Ibid., page 43.

20. Ibid., page 45.

21. Today, two-thirds of French electricity is generated by nuclear power, far more than in other western countries.

22. Interview with Bongo in *Libération,* September 18, 1996.

23. Loïk Le Floch-Prigent, *Affaire Elf, affaire d'état* (Paris: Cherche Midi, 2001), page 54.

24. Before being a resistance fighter, Foccart had also worked with the Vichy government, as did Mitterrand.

25. See, for example, Assemblée Nationale, Rapport d'Information no. 2237, "Sur le contrôle parlementaire des opérations éxterieures," March 8, 2000, available at http://www.assemblee-nationale.fr/rap-info/i2237.asp. "France and Gabon prepare and assure their common defense," the accords say. "Gabon armed forces participate, with the French army, under a single command for the external defense of the community."

26. Robert, *Ministre d'Afrique,* pages 130–131.

27. Ibid., page 211.

28. Bongo, *Blanc comme nègre,* pages 220 and 222.

29. See ibid., page 75.

30. See "Une création de De Gaulle pour contrer l'Amerique," *Libération,* January 13, 2003. The agent was Pierre Guillaumat, who was head of French secret services in London in World War II.

31. From http://www.total-gabon.com. The Elf brand was introduced in 1967, and Elf Gabon got its name in 1971.

32. Pierre Péan, *Affaires africaines* (Paris: Fayard, 1983), page 55.

33. Charles F. Darlington and Alice B. Darlington, *African Betrayal* (New York: D. McKay, 1968).

34. Marc Aicardi de Saint-Paul, *Gabon: The Development of a Nation,* translated by A., F., and T. Palmer (London: Routledge, 1989), page 97.

35. Bongo, *Blanc comme nègre,* pages 76–78.
36. Paul Toungui, subsequently finance minister.
37. Bongo, *Blanc comme nègre,* page 79.
38. Ibid., pages 84–85.
39. Ibid., page 146.
40. Ibid., page 233.
41. Robert, *Ministre d'Afrique,* page 351.
42. Péan, *Affaires africaines,* page 35.
43. Several American presidents were freemasons, along with numerous French politicians active in African affairs.
44. According to some accounts, since renamed *Grande loge symbolique du Gabon.*
45. "L'étrange influence des francs-maçons en Afrique francophone," *Le monde diplomatique,* September 1997. See also "Ouverture vendredi a Libreville de la 6ème Conférence des REHFRAM," *Agence France-Presse,* February 6, 1998. Rival branches of freemasonry caused divisions in French politics too.
46. Author's interview with professor who wishes to remain anonymous, Paris, 2003.
47. Stephen Smith, speaking on BBC, December 1998.
48. World News Connection, August 11, 2005, citing Nku'u Le Messager.
49. Valérie Lecasble and Airy Routier, *Forages en eau profonde: Les secrets de l'affaire Elf* (Paris: Grasset, 1998), page 356.
50. Péan, *Affaires africaines,* page 20, and also see interview with Péan, http://www.voltairenet.org/article7961.html.
51. See Péan, *Affaires africaines,* page 250.
52. Olivier Vallée quoted in "Elf au service de l'état français," *Le monde diplomatique,* April 2000.
53. Germain Mba.
54. See Péan, *Affaires africaines,* first chapter, for a description of Mba's killing.
55. Douglas A. Yates, *The Rentier State in Africa: Oil Rent Dependency and Neocolonialism in the Republic of Gabon* (Trenton, N.J.: Africa World Press, 1996), page 131.
56. For replacing and modernizing Elf's original 40-year lease for an oil permit for 88,000 square kilometers in Gabon. The journalist Péan estimates, on page 29 of his *Affaires africaines,* that Elf was getting a profit margin of $5 a barrel, which is huge in oil industry terms.
57. Banque des états de l'afrique centrale (BEAC) is the regional central bank for six countries in the sub-region.
58. Yates, *The Rentier State in Africa,* page 45, and Darlington and Darlington, *African Betrayal.* This was loosened, but not discarded, in 1972.

59. See, for example, Sanou Mbaye (formerly of the African Development Bank), "France Killing French Africa," January 29, 2004, available at www.taipeitimes.com/News/editorials/archives/2004/01/29/2003096656.

60. See Yates, *The Rentier State in Africa,* page 110.

61. Michel Debré. See Yates, *The Rentier State in Africa,* page 43.

62. As described to me by a South African official who was involved in efforts to build up a rival free zone on São Tomé e Principe.

63. For example, a white man was present at my interview with François Ombanda, CEO of the water and electricity company SEEG, and the representative of the Banque internationale pour le commerce et l'industrie du Gabon (BICIG, Gabon's biggest bank) was French.

64. See Yates, *The Rentier State in Africa,* and IMF, *Direction of Trade Statistics Yearbook* (Washington, D.C.: International Monetary Fund, 2002), page 225, which shows that of Gabon's imports in the first two years of the twenty-first century, two-thirds came from France.

65. According to the Economist Intelligence Unit, June 1999; see "The World's Most Expensive Cities," *BBC News,* available at http://news.bbc.co.uk/1/hi/business/your_money/381351.stm.

66. Marc Aicardi de Saint-Paul, *Gabon: The Development of a Nation,* page 67.

67. Yates, *The Rentier State in Africa,* page 174.

68. According to the Economist Intelligence Unit, Gabon got $35 million in the privatization deal for a 20-year concession.

69. Gabon's national debt is around $4 billion, according to the World Bank's "Gabon at a Glance," September 2002.

70. According to the Economist Intelligence Unit, Gabon Country Report, fourth quarter 2005, "Economic Policy" section, debt service takes up 45 percent of Gabon's 2005 budget and 49 percent of its 2006 budget.

71. The Ivorian president Félix Houphouet-Boigny is credited with coining the term in the 1970s.

72. Antoine Glaser, Stephen Smith, *Comment la France a perdu l'Afrique,* pages 83–85.

Chapter 5—Eva Joly

1. Eva Joly, *Est-ce dans ce monde-là que nous voulons vivre?* (Paris: Arènes, 2003), page 28.

2. Eva Joly, *Notre affaire à tous* (Paris: Arènes, 2000), page 35.

3. From Tim King, "French Favors," *Prospect,* January 2004.

4. Joly, *Notre affaire à tous,* page 59.

5. Ibid., page 116.

6. Ibid., page 120.

7. Ibid., pages 123–126.

8. For a good explanation of the Elf trial, see David Ignatius, "True Crime: The Scent of French Scandal," *Legal Affairs,* May–June 2002, available at http://www.legalaffairs.org/issues/May-June–2002/story_ignatius_mayjun 2002.html.

9. Interview in English on BBC television's *HardTalk* with Tim Sebastian, July 15, 2000, available at http://www.amgot.org/hist/.

10. See, for example, Valérie Lecasble and Airy Routier, *Forages en eau profonde: Les secrets de l'affaire Elf* (Paris: Grasset, 1998), page 52.

11. Ignatius, "True Crime: The Scent of French Scandal."

12. Most commentators speak of the "réseaux Pasqua"; Loïk Le Floch-Prigent, in *Affaire Elf, affaire d'état* (Paris: Cherche Midi, 2001), argues that this was a misnomer: the networks were Pasqua's and Chirac's, together.

13. From "The Power Broker in France's Election: Interior Minister Pasqua Embodies Nation's Social Divide," *San Francisco Chronicle,* April 21, 1995.

14. David Ignatius, "True Crime: The Scent of French Scandal."

15. For example, Loïk Le Floch-Prigent, *Affaire Elf, affaire d'état,* page 22.

16. Interview in English on BBC Television's *Hard Talk with Tim Sebastian,* July 15, 2000.

17. Joly, *Notre affaire à tous,* page III.

18. Joly, *Est-ce dans ce monde-là que nous voulons vivre?* page 36.

19. Ibid., page 43.

20. Ibid., pages 38 and 43.

21. Ibid., page 52.

22. Ibid.

23. Tim King, "French Favors."

24. Loïk Le Floch-Prigent, *Affaire Elf, affaire d'état,* page 24, and Joly, *Est-ce dans ce monde-là que nous voulons vivre?* page 130.

25. Antoine Gaudino, quoted in Tim King, "French Favors," *Prospect,* January 2004.

26. Joly, *Est-ce dans ce monde-là que nous voulons vivre?* page 57.

27. See Antoine Glaser and Stephen Smith, *Ces messieurs Afrique 2: Des réseaux aux lobbies* (Paris: Calmann-Lévy, 1997), page 193, and Valérie Lecasble and Airy Routier, *Forages en eau profonde,* page 147.

28. Rassemblement pour la République.

29. Lecasble and Routier, *Forages en eau profonde.*

30. See, for example, Global Witness's *Time for Transparency,* March 24, 2004, page 38.

31. Author's interview with Verschave, Berlin, June 2003.

32. Lecasble and Routier, *Forages en eau profonde,* page 252.

33. Bongo freely admits his and his family's shareholdings in Fiba, as he did, for example, in "Ma vérité sur l'affaire Elf," a *Jeune Afrique* interview of May 7, 2003. According to Olivier Vallée's "Elf au service de l'état

français," *Le monde diplomatique,* April 2000, the shareholding of Fiba includes 43 percent by Elf, 35 percent by Bongo's family, and another 16 percent by other Gabonese "private interests." This is also referred to in the Elf indictment.

34. Loïk Le Floch-Prigent, *Affaire Elf, affaire d'état,* page 104. A schematic of some of the structures set up involving Elf Gabon is also provided in a presentation by Eva Joly at http://www.iadb.org/etica/Documentos/uru_jol_comba-i.ppt.

35. In his book *Affaire Elf, affaire d'état,* pages 102–108, Le Floch describes the operations of Fiba in detail. See also, for example, Glaser and Smith, *Ces messieurs Afrique 2,* page 118.

36. By 1975, Elf Gabon was already producing around 65 million barrels per year. See http://www.total-gabon.com/stat/production.htm. Note that this is operated, not equity, production.

37. For example, the arms dealer Jacques Monsieur. See *The Field Marshal,* Center for Public Integrity, November 15, 2002.

38. Global Witness, *Time for Transparency.*

39. Speaking on the BBC's "Gabon: The Oil-Rigged State," December 23, 1998.

40. Loïk Le Floch-Prigent, *Affaire Elf, affaire d'état,* pages 35 and 55. These are the figures given by Le Floch: official commissions of 800 million French francs, on a turnover of 200 billion francs per year. According to Lecasble and Routier, *Forages en eau profonde,* page 266, the 800 million francs had risen from 100 million francs per year in the time of Pierre Guillaumat, the first Elf president, to 200 million in the time of Guillaumat's successor Albin Chalandon, to 300 million in the time of Chalandon's successor Michel Pecqueur. Added to the official commissions of 800 million francs should be added "occult" payments, bringing the annual total to around 1.5 billion francs.

41. Maurice Robert, *Ministre d'Afrique: Entretiens avec André Renault* (Paris: Seuil, 2004), page 273. Robert is a former French secret service member and former French ambassador in Libreville.

42. Joly, *Est-ce dans ce monde-là que nous voulons vivre?* pages 87–88.

43. Ibid., page 15.

44. Though the U.S. Foreign Corrupt Practices Act of 1977 did outlaw bribery, and this was followed up by the OECD Convention on Combating Bribery of Foreign Officials in International Business Transactions, which entered into force in 1999, France ratified the convention only three years later (and the United Kingdom, disgracefully, ratified it only in 2002).

45. Joly, *Est-ce dans ce monde-là que nous voulons vivre?* page 170.

46. "Gabon—Digging Deeper Holes," *Africa Confidential,* May 11, 1999.

47. For example, *Jeune Afrique,* June 2005, outlined a visit to Europe during which Bongo met Gerhard Schroeder, José Manual Durão Barroso, Jacques Chirac, Dominique de Villepin, and Nicolas Sarkozy, among others.

48. In the Economist Intelligence Unit's 2002 Gabon country profile, net French development assistance was minus $14.5 million in 2000. See also the reference tables "Net Official Development Assistance" in the Economist Intelligence Unit, Gabon Country Profiles, 2000 and 2003.

49. See http://www.sysmin-gabon.org/index.php?page=sysmin-gab.html.

50. Total Gabon annual report, http://www.total-gabon.com/documents/2005/totalgabon2005/index.htm.

51. Joly, *Est-ce dans ce monde-là que nous voulons vivre?*

52. Eric Decouty, "Les juges financiers baissend pavillon," *Le Figaro,* March 3, 2006.

53. Joly, *Notre affaire à tous,* page 243.

54. See Senate Hearings, November 9, 1999, page 589, which describes him as "Nadhmi Auchi, who is close to Saddam Hussein."

55. Senate Hearings, page 69. The Foreign Corrupt Practices Act, for example, outlaws foreign bribery by American-listed companies, but it does not outlaw looting of state treasuries by foreign officials.

56. Senate Hearings, page 863.

57. Ibid., pages 569–571.

58. Ibid., page 512.

59. Omar Bongo, *Blanc comme nègre: Entretiens avec Airy Routier* (Paris: Grasset, 2001), pages 285–291. This is a condensed version of what he said.

60. Senate Hearings, page 567.

61. Two Citibank officials who had been involved with the Bongo accounts.

62. Senate Hearings, page 147.

63. Ibid., page 75.

64. Testimony by Raymond Baker, Senate Hearings, pages 85–86.

65. http://www.transparency.org/integrityaward/winners/winners_2001.html.

66. See http://www.bloomberg.com/apps/news?pid=20601085&sid=aH.R_6Jady7c&refer=europe.

67. http://www.ft.com/cms/s/7a5dbb1c–014a–11db-af16–0000779e2340.htm.

68. See http://www.ft.com/cms/s/f61d9c12–1816–11db-b198–0000779e2340.html and http://www.ft.com/cms/s/cc4bcda2–20bb–11db–8b3e–0000779e2340.html.

69. Widely reported. The first article about this was "La justice française s'intéresse à l'ancienne société de Dick Cheney," *Le Figaro,* June 2, 2003.

70. See "Out of arms way," *The Guardian* (UK), August 8, 2003.

71. See http://www.guardian.co.uk/armstrade/story/0,1014976,00.html.

72. Joly, *Notre affaire à tous,* page 20.

73. U.N. Human Development Report, 2005.

74. Data from World Bank, "Gabon at a Glance," September 2002; U.N. Human Development Report, 2002; and IMF, "Gabon Report," February 5, 2005.

75. The U.N. Human Development Report for 2004 gives Gabon a growth rate of minus 0.1 percent per year from 1990 to 2001.

76. U.N. Human Development Report, 2003.

77. The Franc Zone's dedicated Web site, http://www.izf.net, reports that of Libreville's 190,000 tonnes of food, 60 percent is imported; for animal products, the figure is 70 percent.

78. Bongo, *Blanc comme nègre,* pages 177–178.

79. See, for example, IMF, "Article IV Consultation—Staff Report," June 2006, page 10.

Chapter 6—André Milongo

1. Widely reported. See, for example, "Natural Resources and Violent Conflict," paper by Ian Bannon and Paul Collier, World Bank, 2003, or *Bottom of the Barrel: Africa's Oil Boom and the Poor,* by Catholic Relief Services, June 2003.

2. In François-Xavier Verschave, *L'envers de la dette* (Paris: Agone, 2002).

3. As revealed in, and widely reported from, the Elf trials. See, for example, Karl Laske, "La pompe Afrique: Tours de passe-passe," *Libération,* January 13, 2003.

4. Thomas Pakenham, *The Scramble for Africa,* (London: Abacus, 1992) page xxiv.

5. Ibid., page 358.

6. Ibid., page 359.

7. Ibid., page 154.

8. From a BBC chronology, February 27, 2004.

9. From Adam Hochschild's classic *King Leopold's Ghost* (London: Papermac, 2000).

10. Douglas A. Yates, *The Rentier State in Africa: Oil Rent Dependency and Neocolonialism in the Republic of Gabon* (Trenton, N.J.: Africa World Press, 1996), page 90.

11. Centre des Archives d'Outre-Mer, Aix-en-Provence, France.

12. John F. Clark, "The Neo-colonial Context of the Democratic Experiment of Congo-Brazzaville," *African Affairs,* April 2002.

13. Fiba has been replaced by BGFIbank, using the same premises. The old Fiba system has been dismantled.

14. Valérie Lecasble and Airy Routier, *Forages en eau profonde: Les secrets de l'affaire Elf* (Paris: Grasset, 1998), page 237.

15. See "Ce qui dit l'ordonnance de renvoi." At the subsequent Elf trials in Paris, officials in court said that the head of Elf financed an attempted coup d'état on January 15, 1992, by the chief of army staff to get rid of Milongo.

16. Jacques Sigolet, director of Elf's Fiba bank, in Elf indictment, quoted in Global Witness, *Time for Transparency,* page 21.
17. In Olivier Vallée, "Les cycles de la dette," *Politique africaine,* October 1988, page 3.
18. Verschave, *L'envers de la dette,* page 24; Olivier Vallée, "Les cycles de la dette"; and IMF and World Bank reports, quoted in Global Witness's *Time for Transparency,* March 24, 2004, page 18.
19. Vallée, "Les cycles de la dette," pages 17–19.
20. In the words of Jacques Sigolet, from Elf indictment, cited in Global Witness, *Time for Transparency,* page 20. Sigolet said that Elf would set up a company in, say, Switzerland, which would lend at a higher interest rate to a bank, which would then lend at an even higher rate to Congo.
21. The rescheduling deal, and its link to the award to Elf of the giant Nkossa field and others, was described in court by Lissouba's finance minister and by Jacques Sigolet.
22. See, for example, Verschave, *L'envers de la dette,* page 110, about the CIAN.
23. Global Witness, *Time for Transparency,* page 21.
24. See African Affairs, Volume 101, April 2002, page 181.
25. Loïk Le Floch-Prigent, quoted in *L'Express,* December 12, 1996.
26. From court testimony of André Tarallo and others in Paris; see www.rfi.fr/fichiers/evenements/elf/circuits.asp: "Bongo asked Sassou to make a deal with Lissouba, and Bongo asked Elf to support Lissouba's campaign."
27. Author's interview with Patrice Yengo, Paris.
28. See African Affairs, Volume 101, April 2002, page 182.
29. In fact, Le Floch-Prigent testified later in Paris that he tried to get this money delayed, "so they could not pay the electorate with this money, and the elections would take place in peace." See the RFI document.
30. This deal is well documented in Congo and the French media. Lissouba and four of his ministers were eventually convicted in absentia by a Brazzaville tribunal in which these terms of the deal were explicitly stated. See, for example, "L'ancien président Pascal Lissouba condamné par contumace à 30 ans de travaux forcés pour haute traison," *Associated Press,* December 28, 2001.
31. "Congo: Pour quelques dollars de plus," *La lettre du continent,* May 27, 1993. See also Verschave, *L'envers de la dette,* page 45.
32. See the RFI document. "Je prends un coup de sang, j'insulte les gens d'Oxy que je connais bien."
33. Economist Intelligence Unit, Congo Country Report, first quarter 1996.
34. Author's interview with Lambert Galibali, Paris, May 2003.
35. Janet MacGaffey and Rémy Bazenguissa-Ganga, *Congo-Paris: Transnational Traders on the Margins of the Law* (Oxford: James Currey, 2000).

36. Economist Intelligence Unit, Country Report, second quarter 1996. The Chevron signature for the Marine IV license, and the Gore visit, happened on December 4, 1995.

37. See http://query.nytimes.com/gst/fullpage.html?res=9A02E3D91439F932 A05754C0A960958260.

38. Later, at the Elf trial, Elf officials confirmed the supply of weapons to Sassou via Gabon. See, for example, Verschave, *L'envers de la dette,* page 33. According to *Le canard enchaîné* and the magistrates' document, reprinted by RFI, the arms flew from Le Bourget airport in Pairs to Gabon, then moved from Franceville in eastern Gabon to Oyo, Sassou's birthplace.

39. Norbert Dabira, *Brazzaville à feu et à sang* (Paris: Harmattan, 1998), page 15.

40. Interview with Lissouba in the *New African,* May 1998.

41. See "Congo: Le nerf de la guerre," *La lettre du continent,* October 2, 1997; repeated in Verschave, *L'envers de la dette.* Elf briefly halted its royalty payments to Lissouba, but later resumed them, under pressure from Socialists in Paris. See Economist Intelligence Unit, Congo Country Report, first quarter 1998.

42. Recorded by the BBC in "Gabon: The Oil-Rigged State," December 23, 1998; part of a two-series program called "A Mission to Civilize."

43. "Paris a choisi le vainqueur au Congo," *Le canard enchaîné,* October 22, 1997.

44. Lambert Galibali, whom author interviewed in May 2003 in Paris.

45. Economist Intelligence Unit, Congo Country Report, fourth quarter 1997.

46. Economist Intelligence Unit, Congo Country Report, first quarter 1998; interview with U.N. official in Congo in 2002.

47. Economist Intelligence Unit, Brazzaville Country Report, third quarter 1997.

48. "Soupçonné d'avoir perçu 5 millions de dollars, Jean-Charles Marchiani (RPF) a été mis en examen dans l'affaire Elf," *Le monde,* January 31, 2004. Another warlord, Bernard Kolelas, reportedly received $2.4 million from Elf's Swiss account, Rivunion. See also Renaud Lecadre, "Affaire Elf, quand c'est fini, il y en a encore," *Libération,* July 10, 2003 and see http://www.publicintegrity.org/bow/report.aspx?aid=155.

49. Elf indictment, page 91, and see Global Witness, *Time for Transparency,* page 22, for detailed description.

50. Verschave, *L'envers de la dette,* page 84, and see Global Witness, *Time for Transparency,* page 22, which describes the deal in detail.

51. Loïk Le Floch-Prigent, *Affaire Elf, affaire d'état* (Paris: Cherche Midi, 2001), page 108.

52. Rough estimate given me by Bill Paton, the top U.N. representative in Congo, in May 2002.

53. U.S. State Department, Country Report on the Republic of Congo, 2000, and Global Witness, *Time for Transparency*, page 24; a similar estimate was given me by Bill Paton, the top U.N. representative in Congo, in May 2002.

54. Antoine Glaser and Stephen Smith, *Ces messieurs Afrique 2: Des réseaux aux lobbies* (Paris: Calmann-Lévy, 1997).

55. Ibid.

56. Nominal and net present value; from IMF, http://www.imf.org/external/pubs/ft/scr/2005/cr05391.pdf, November 2005, pages 13 and 14. Data includes $1 billion owed by the state oil company. A subsequent IMF deal cut this to under $7 billion, and high oil prices have reduced a bit of pressure. See http://www.imf.org/external/pubs/ft/scr/2005/cr0539.pdf and http://www.imf.org/external/pubs/cat/longres.cfm?sk=19451.0.

57. Several are recorded in 2002 onward in Global Witness, *Time for Transparency*, page 31.

58. From accounting sources in Brazzaville. Some of this is known as "prefinancing." Global Witness has cited annual interest rates of up to 170 percent annually. (See Global Witness, *The Riddle of the Sphynx: Where Has Congo's Oil Money Gone?* December 2005.)

59. See http://rightweb.irc-online.org/profile/982. Also see "Congo Battle Looms over White House," *New York Sun*, June 5, 2006.

60. The London High Court in 2005 found that nearly $500 million of Congo's oil money passed through fake private companies controlled by the chairman of Congo's state oil company, using tax havens. Court documents in New York describe a $650 million loan to Congo being covered with $1.4 billion in oil cargoes.

61. Denis Sassou Nguesso, "L'Afrique, le Congo, et lui," *Jeune Afrique*, February 19–25, 2006.

62. Global Witness, *Time for Transparency*, page 31.

63. Letter was in 2002. From Global Witness, *Time for Transparency*, page 19.

64. From sources in Brazzaville, November 2003; also described in Global Witness, *Time for Transparency*, page 33.

65. For a good description of this, see *Petroleum and Politics in the Gulf of Guinea* by Ricardo Soares, Cambridge University (U.K.), May 2005, Chapter 2.

66. Médécins sans frontières—Holland.

67. According to a doctor from Médécins sans frontières, in October 2003.

68. See, for example, Global Witness, Time for Transparency, page 25, and "Cleaned out," Africa Confidential, December 5, 2003.

Chapter 7—Obiang Nguema

1. Quoted in the International Consortium of Investigative Journalists' "The Curious Bonds of Oil Diplomacy," November 6, 2002.

2. From IMF Country Report 03/386, December 2003, pages 37 and 45. The IMF puts oil exports at 1,241 bn CFA in 2001, and government oil revenue at just 286 bn CFA.

3. From IMF Country Report no. 99/113, page 21.

4. Planning Minister Fortunato Ofa Mbo, in "Informe sobre economia y desarollo," Bata, July 4–6, 2001, recorded agriculture as constituting 2.6 percent of GDP.

5. Fernando Abaga, "Las consequencias socio-economicas del petroleo en guinea ecuatorial: Del 'boom' a la quiebra," May 1999.

6. Jean-François Bayart, Stephen Ellis, and Béatrice Hibou, *The Criminalization of the State in Africa,* translated by Stephen Ellis (Oxford: James Currey, 1999), page 26.

7. Amnesty International, *Equatorial Guinea: A Country Subjected to Terror and Harassment,* January 1, 1999, available at http://web.amnesty.org/library/Index/ENGAFR240011999?open&of=ENG–380.

8. Asodegue.

9. "Equatorial Guinea's 'God,' "BBC News, July 26, 2003, available at http://news.bbc.co.uk/1/hi/world/africa/3098007.stm.

10. In response to author's question in Bata, August 2001, but he echoed the words in a speech to U.S. corporate council on Africa on February 8, 2002.

11. Ken Silverstein, "Oil Boom Enriches African Ruler," *Los Angeles Times,* January 20, 2003.

12. "U.S. Senate Hearing before the Permanent Subcommittee on Investigations," July 15, 2004, pages 85 and 105.

13. See http://www.washingtonpost.com/wp-dyn/articles/A28396–2004May14.html or http://www.washingtonpost.com/wp-dyn/articles/A58805–2004May26.html.

14. Senate Hearing, page 605.

15. Ibid., pages 2–3. Money was traced flowing from Riggs to the associates of two of the hijackers. The Senate also probed accounts belonging to Augusto Pinochet, the former Chilean ruler.

16. Senate Hearing.

17. Ibid., page 45.

18. Ibid., page 85.

19. From author's inquiries in Washington, November 2005.

20. Senate Hearing, pages 2 and 5–6.

21. Ibid., page 307.

22. Ibid., page 165.

23. Ibid., pages 130–132.

24. Ibid., pages 214 and 215 and http://www.guardian.co.uk/equatorialguinea/story/0,1497227,00.html.

25. Senate Hearing, page 157.

26. Ibid., pages 14 and 161.

27. "Who Would You Say Is the World's Worst Dictator?" *Parade*, February 22, 2004. Obiang was number 6.

28. Senate Hearing, page 296.

29. Ibid., pages 270–295.

30. Ibid., page 130.

31. Ibid., page 29.

32. Author's interview, in the presence of the Jeune Afrique journalist Jean-Dominique Geslin, November 2003.

33. Ken Silverstein, "U.S. Oil Politics in the 'Kuwait of Africa,'" *Nation*, April 4, 2002, available at http://www.thenation.com/doc/20020422/silverstein. This was before the Riggs scandal broke.

34. Senate Hearing, page 329.

35. Nusiteles G.E., a telecommunications venture. Senate Hearing, pages 16 and 164.

36. Senate Hearing, page 7, and 71–72.

37. Ibid., page 69 and 200.

38. "Marathon & GEPetrol Finalize Equatorial Guinea LNG Project," Marathon Oil press release, June 22, 2004, and Senate Hearing, pages 164 and 647.

39. Senate Hearing, page 221.

40. From author's conversation with Senate staffer, November 2005.

41. Senate Hearing, page 63.

42. Ibid., page 72, for Amerada Hess, page 834 for ExxonMobil, and page 870 for Marathon.

43. Ibid., page 68.

44. See U.S. Department of Justice, "Riggs Bank Sentenced to Pay $16 Million Fine for Criminal Violation of the Bank Secrecy Act," March 29, 2005.

45. U.K. Channel Four News interview with President Obiang, broadcast November 18, 2003.

46. "Les hommes d'affaires en Guinée Equatoriale à l'origine de la corruption (Obiang)," *Agence France-Presse*, October 10, 2003. Global Witness reproduced this in their report *Time for Transparency*, March 24, 2004.

47. Global Witness, *Time for Transparency*, page 56.

48. *African Energy*, Issue 57, December 2002, http://www.africa-energy.com/html/public/demo/analysis/analysis1.html.

49. "Equatorial Guinea: A Better Image for the Latest Kuwait of Africa," *Africa News Service*, May 26, 1999.

50. Corporate Council on Africa, *Equatorial Guinea: A Country Profile for U.S. Businesses*, April 2001.

51. See Foreign Agents Registration Unit (FARA) on U.S. Department of Justice Web site, http://www.udsoj.gov/criminal/fara.

52. See, for example, http://www.gasandoil.com/goc/company/cnn84765.htm.

53. See Ken Silverstein, "Our New Favorite Despot," Salon.com, April 29, 2002, and many other articles, including Ken Silverstein, "U.S. Oil Politics in the 'Kuwait of Africa,'" http://www.altassets.net/news/arc/2005/nz7831.php, http://www.icij.net/about/release.aspx?aid=14, and http://www.crimelibrary.com/gangsters_outlaws/cops_others/thatcher_and_mann/16.html.

54. See Ken Silverstein, "The Crude Politics of Trading Oil," *Los Angeles Times*, December 12, 2002, and http://www.publicintegrity.org/about/release.aspx?aid=21.

55. Quote from Ken Silverstein, "U.S. Oil Politics in the 'Kuwait of Africa.'"

56. See http://www.nci.org/05nci/02/IPC-feb7.htm, or http://www.iranpolicy.org/index.php?option=com_content&task=view&id=19&Itemid=29.

57. Senate Hearing, page 593.

58. The joint venture was called Nusiteles; Senate Hearing, pages 302, 346–356, and 693–695.

59. McColm admitted that IDS and IFES received Mobil money, in an e-mail to journalist David Hecht, July 2, 1999. CMS, which sold out to Marathon, also provided support.

60. See, for example, Duncan Campbell, "Marketing the New 'Dogs of War,'" *U.S. Center for Public Integrity*, October 30, 2002, available at http://www.publicintegrity.org/bow/report.aspx?aid=149.

61. Two sources who attended the meeting differ as to whether a coup plot was explicitly predicted or not.

62. "It is potentially a very lucrative game," begins the document assessing threats. "We should expect bad behavior; disloyalty; rampant individual greed; irrational behavior (kids in toyshop type); back-stabbing . . . and similar ungentlemanly activities."

63. A transcript of this emerged: Claim no. HQ04X02003 in the High Court of Justice, Queen's Bench Division, between (1) President Obiang and (2) Equatorial Guinea, and (1) Logo Ltd. (British Virgin Islands), (2) Systems Design Ltd. (Bahamas), (3) Greg Wales, (4) Simon Francis Mann, (5) Ely Calil, and (6) Severo Moto, March 2004.

64. Archer's name appears in several stories, such as Barbara Jones, "African Leader: Arrest Mark Thatcher over Coup Plot," *Mail on Sunday* (U.K.), August 15, 2004.

65. In Jannuary 2005, Thatcher admitted an "unwitting" role in financing the plot in a plea bargain that saved him from jail. He was fined and received a suspended prison sentence.

66. This was widely quoted. See, for example, "Thatcher and a Very African Coup," *Guardian* (U.K.), August 26, 2004, available at http://www.guardian.co.uk/equatorialguinea/story/0,15013,1291453,00.html.

67. Adam Roberts, *The Wonga Coup: The British Mercenary Plot to Seize Oil Billions in Africa* (London: Profile Books, 2006), page 63.

68. See http://observer.guardian.co.uk/politics/story/0,6903,1361298,00.html.

69. See Roberts, *The Wonga Coup*, pages 79, 81, 84, 134, 136, 139, 140, 141, 143, 166–167, 174, 184–5, 191.

70. See, for example, "Equatorial Guinea: Ripe for a Coup," *BBC News*, March 11, 2004, available at http://news.bbc.co.uk/1/hi/world/africa/3500832.stm, or "Exiled leader in Spain denies any coup attempt," *The Guardian*, August 27, 2004.

71. See, for example, "Mark Thatcher's Dangerous Connections," *Vanity Fair*, December 20, 2000, available at http://www.vanityfair.com/commentary/content/articles/041220roco01?page=6, or "What Made Jack Straw Tell the Truth about the Botched Coup in Equatorial Guinea," *Spectator*, November 27, 2004.

72. See "How much did Straw know and when did he know it?," *The Observer*, November 28, 2004

73. This first quote is from http://washingtontimes.com/upi-breaking/200409 24–052440–8265r.htm.

74. From author's recording of Obiang speech, via telephone from Malabo. Also see "Equatorial Guinea foils 'plot,'" *Reuters*, March 10, 2004. He was speaking in Spanish.

75. See "Zimbabwe: West aided Mercenaries," *Reuters*, March 10, 2004

76. http://www.guardian.co.uk/international/story/0,3604,1289837,00.html.

77. TNO reportedly had a distribution contract with Time Warner, according to the due diligence Riggs carried out on Teodorín, Senate Hearing, page 1298.

78. Senate Hearing, page 1298.

79. Author's interview with Teodorín, Bata, July 2001.

80. Press Release issued by M. Teodoro Nguema Obiang, Minister of State, Government of the Republic of Equatorial Guinea, June 13, 2001.

81. Author's interview with a western ambassador, Malabo, July 5, 2001.

82. See IMF, "Equatorial Guinea: Selected Issues and Appendix," June 2006, pages 100 and 104. Government oil and gas revenue was CFA 766.3 billion, worth $1.17 billion, while total petroleum exports were worth $4.51 billion. ExxonMobil's Zafiro costs were paid off in 2004; in 2005 the ratio of government revenue to total exports rose to just 31 percent.

83. The IMF, in its 2004 publication *Lifting the Oil Curse: Improving Petroleum Revenue Management in Sub-Saharan Africa*, page 5, estimated that governments in the region collected an average 50 percent of the value of their oil in 2001, ranging from 90 percent in Nigeria to 21 percent for Equatorial Guinea.

84. Britain's BG group signed a letter of understanding in May 2003 for a 17-year gas supply agreement with Equatorial Guinea LNG holdings Ltd, but

the purchase agreement was only signed in June 2004, after the plot. The author has no indication that this contract is abnormally profitable.

85. See IMF, "Equatorial Guinea: Selected Issues and Appendix," June 2006, page 37.

86. U.K. Channel Four News interview with President Obiang, broadcast November 18, 2003, reproduced in Global Witness's *Time for Transparency*.

87. See http://www.imf.org/external/np/sec/pn/2005/pn0561.htm.

88. "World Bank President Paul Wolfowitz' Remarks at the Corporate Council on Africa Dinner," Thursday, June 23, 2005, transcript by the *Federal News Service*, Washington, D.C; available at http://www.africacncl.org/(zlck15ynuec24e45ffbv2qeq)/Default.aspx.

89. See Office of Senator Joe Biden, U.S. Congress, "Equatorial Guinea: Letter to President Bush on Decisions Regarding Military Training Program and U.S. AID Assistance," May 19, 2006.

90. U.S. State Department, "Remarks with Equatorial Guinean President Teodoro Obiang Nguema Mbasogo before Their Meeting, Secretary Condoleezza Rice," April 12, 2006, available at http://www.state.gov/secretary/rm/2006/64434.htm.

91. http://www.whitehouse.gov/news/releases/2004/01/20040112-3.html.

92. http://guinea-equatorial.com/news.asp?DocID=57.

93. See "Hausmitteilung," *der Spiegel* 35/2006, 28th August 2006, page 3, and English translation, "Torture and Poverty in Equatorial Guinea," by Alexander Smlotczyk, on http://service.spiegel.de/cache/international/spiegel/0,1518,434691,00.html.

94. "Marathon Oil Corporation at Banc of America Securities Energy Conference—Final," *FD (Fair Disclosure) Wire*, November 29, 2005.

95. CCA document, April 2001, page 40.

Chapter 8—Fradique de Menezes

1. Alex Newton, *Lonely Planet: Central Africa* (Oakland, Calif.: Lonely Planet, 1994).

2. See, for example, Speech by H. E. Fradique de Menezes at Closing Plenary Lunch with Hon. Colin Powell, U.S. Secretary of State, Corporate Council on Africa, June 27, 2003.

3. From Fradique's speech at Closing Plenary Lunch with Colin Powell, Corporate Council on Africa, June 27, 2003.

4. Under São Tomé's constitution, the president has executive powers only over foreign affairs and defense. Economy (oil) is the exclusive preserve of government; all oil deals since 1997 have been signed by government ministers.

5. Gerhard Seibert, *Comrades, Clients, and Cousins: Colonialism, Socialism, and Democratization in São Tomé and Príncipe* (Leiden, Netherlands: Research

School of Asian, African, and Amerindian Studies, Leiden University, 1999), page 19, and author's subsequent correspondence with Seibert.

6. William A. Cadbury, *Labor in Portuguese West Africa* (London, 1910), quoted by Seibert, *Comrades, Clients, and Cousins,* page 19.

7. According to Seibert, these families are Pinto da Costa, Trovoada, Daio, Tiny, Espirito Santo, D'alva, Daio, Costa Alegre. Fradique's is not in here.

8. See "Management of Wealth under the Permanent Income Hypothesis: the case of Sao Tome e Principe," IMF, July 1, 2006, page 5. Note that life expectancy in African island states tends to be higher than on the continent.

9. Correspondence with São Tomé expert Gerhard Seibert.

10. Author's interview, 2002.

11. For example, $110 million was offered (but not all disbursed) for the 1986–1990 development program.

12. President Trovoada, who took over from Pinto da Costa.

13. Information provided by Ken Silverstein and Gerhard Seibert.

14. Seibert, *Comrades, Clients, and Cousins,* page 163.

15. Ibid., pages 208–209.

16. "We Were Sold into Porn Slavery, Cry African Islands," *Register,* December 20, 2004, available at http://www.theregister.co.uk/2004/12/20/sao_tome_denounces_own_domain/.

17. Seibert, *Comrades, Clients, and Cousins,* page 241.

18. IMF Report for São Tomé for 1999 Article IV Consultation, May 2000, page 6.

19. Ibid., page 37. Of the 84,499 million dobra public investment budget, foreign donors financed 82,028 million.

20. Jedrzej Georg Frynas, Geoffrey Wood, and R. M. S. Soares de Oliveira.

21. According to Seibert, Trovoada also faced fierce opposition to recognizing Taiwan from the government of Raúl Bragança. The stalemate ended only in October of that year, thanks to Taiwanese checkbook diplomacy.

22. See Ken Evans's statement in the Pedro Motú chapter.

23. Author's interview, October 15, 2003.

24. "Technical Assistance Agreement by and among São Tomé Principe . . . and Mobil Exploration and Producing Services Inc.," September 10, 1998. The after-royalty profit share for Mobil would be 100 percent for 0–50,000 barrels per day, sliding up to just 50 percent for production in excess of 250,000 barrels per day. "Mobil may not select more than 22 blocks in total."

25. See, for example, Office of the Attorney General, São Tomé and Principe, Investigation and Review, Second Bid Round, Joint Development Zone Nigeria and São Tomé and Principe, December 2, 2005, page 6.

26. Ken Silverstein, author of "Sinking Its Hopes into a Tiny Nation," *Los Angeles Times,* May 24, 2003, filed this in an original version of his story. With kind permission.

27. ERHC annual report, SEC form 10KSB, December 31, 2001.

28. According to one source, the Trovoadas were also instrumental in getting Offor involved.

29. JDA official Hassan Tukur made a presentation on this during a JDA road show.

30. Office of the Attorney General, São Tomé and Principe, Investigation and Review, Second Bid Round, Joint Development Zone Nigeria and São Tomé and Principe, December 2, 2005, page 4.

31. The minister was Dubem Oniya. The renegotiated contract has been filed at http://www.sec.gov/Archives/edgar/data/799235/000091205701518663/a2 051176zex–1.txt.

32. IMF, "2001 Article IV Consultation and Staff-Monitored Program," February 2002, page 10.

33. Jedrzej Georg Frynas, Geoffrey Wood, and R. M. S. Soares de Oliveira, "Business and Politics in São Tomé e Principe: From Cocoa Monoculture to Petro-State," *Royal African Society*, 2003, page 12: "We are not aware . . . of any similar precedent in the history of Africa's oil industry, since the end of colonialism."

34. IMF, "São Tomé—Letter of Intent, and Technical Memorandum of Understanding," January 9, 2002, on the IMF Web site.

35. Later he stated publicly that Offor's money was for his election campaign.

36. ERHC ceded significant royalty interests and other payments; its combined 30 percent options spread across two licenses was expanded to 125 percent spread across six, and most of their obligations to pay up-front signature bonuses were waived.

37. This was the loss just for the waived signature bonuses. Office of the Attorney General, São Tomé and Principe, Investigation and Review, Second Bid Round, Joint Development Zone Nigeria and São Tomé and Principe, December 2, 2005, page 64.

38. Fradique, in return, accused the authors of the letter of corruption, but was then forced, under threat of legal action, to retract the accusation.

39. First as exiled opposition figures in Gabon, then in the apartheid-era South African Buffalo Battalion, then with Executive Outcomes.

40. Also a former foreign minister; now ambassador in Abuja.

41. These examples were not directly related to oil, but were symptoms of the extremely fraught atmosphere stirred up by oil fever.

42. Office of the Attorney General, São Tomé and Principe, Investigation and Review, Second Bid Round, Joint Development Zone Nigeria and São Tomé and Principe, December 2, 2005, page 25.

43. Ibid., Appendix E, page 2. The Appendix is the report of São Tomé's National Petroleum Agency.

44. Securities Exchange Commission, Form 8-K, May 4, 2006.

45. http://www.harpers.org/sb-soa-tome–28383298.html.
46. See "São Tomé and Príncipe Enacts Model Oil Revenue Management Law," http://www.earthinstitute.columbia.edu/news/2005/story01–07–05.html.
47. Chad, for example, set up a savings fund with World Bank help, but some of the proceeds that were supposed to be spent on development were instead spent on buying arms; the government pushed in 2005 for the fund mechanism to be renegotiated and weakened, prompting an outraged response from the World Bank.

Chapter 9—Arcadi Gaydamak

1. "Signature bonuses" are up-front nonrecoverable down payments, and are not illegal.
2. See Yossi Melman, "He's Just a Big Shot," *Haaretz,* January 13, 2002.
3. See Global Witness, "All the President's Men: the devastating story of oil and banking in Angola's privatised war," March 2002, page 13.
4. See Yossi Melman, "Me? An Oligarch?" *Haaretz,* March 18, 2005.
5. See, for example, Yossi Melman and Julio Godoy, "The Influence Peddlers," *U.S. Center for Public Integrity,* November 13, 2002, available at http://www.publicintegrity.org/bow/report.aspx?aid=154.
6. This is covered quite widely in the French press, which quotes Curial's court testimony.
7. See "Un témoignage éclaire les dessons des ventes l'armes à l'Angola," *Le Monde,* April 3, 2003.
8. The mercenary force-multiplier input of Executive Outcomes was probably more important than the arms deals.
9. The purchase happened after my interview, so I did not ask him about it.
10. See Raymond W. Baker, *Capitalism's Achilles Heel: Dirty Money and How to Renew the Free-Market System* (Hoboken, N.J.: Wiley, 2005), pages 159–160.
11. See, for example, "Poll Rocks the House That Jacques Built," *Observer* (U.K.), June 20, 1999.
12. Widely reported. See, for example, http://energycommerce.house.gov/108/News/05122005_1522.htm.
13. They also arrested Jean-Christophe Mitterrand, the former president's son who had also helped with the Angola deals, turning a colorful judicial matter into an international scandal.
14. Quite widely reported. See, for example, "The Arms Dealer Next Door: International billionaire, French prisoner, Angolan weapons broker," *Arizona Republican.* "Who is Pierre Falcone?" by Ken Silverstein, *In These Times,* December 22, 2001.
15. See, for example, "Une enquête sur une société de vente d'armes vise des personnalités politiques," *Le monde,* December 8, 2000.

16. Arcadi's figure of $6 billion fits with research that I carried out with the help of the *Financial Times* in 2002, this debt was estimated at $6.25 billion; this figure was published in the "Foreign Trade and Payments" section of the Economist Intelligence Unit's Country Report for Angola in February 2002. See IMF, "Angola: Recent Economic Developments," August 2000, page 9, which estimates Angola's external debt in 1995 at $11.7 billion.

17. There is some discussion of this in the executive summary to KPMG's "oil diagnostic" report on Angola's oil sector in 2003.

18. It should be noted that IMF data from the time put the debt at $125 billion: see IMF, "External Borrowing by the Baltics, Russia, and Other Countries of the Former Soviet Union: Developments and Policy Issues," June 1997.

19. Sure enough, IMF data shows Angola's debt falling by more than $3 billion from 1995 to 1996, from $11.7 billion to $8.5 billion. IMF, "Angola: Recent Economic Developments," August 2000, page 9.

20. It is not possible to calculate this exactly, since the data is opaque. On a debt of $11.675 billion in 1995, a cut of one percentage point in the interest rate would yield over $100 million per year. However, only a portion of this debt—perhaps a quarter—would realistically have been paid back under normal circumstances, so the saving would have been less than this.

21. He said that the market value of the notes was 5 to 15 percent of their $1.5 billion face value, but he bought them for 50 percent, or three-quarters of a billion.

22. That Angola agreed to pay 100 percent of face value has not been publicly admitted, although the Angolan private newspaper *Semanário angolense,* in an article on April 6, 2004, stated that under the agreement signed in May 1997, Sonangol would pay $290.3 million plus five payments of $242 million between November 30, 1997, and December 31, 2004, to redeem the Russian debt. In addition, in an article entitled "Le règlement de la dette angolaise aurait donné lieu à des détournements de fonds," France's *Le monde* stated that "le 30 mai 1997, Abalone signe un accord de livraison de pétrole pour une valeur de 1,5 milliard de dollars avec la compagnie Sonangol, agissant au nom du gouvernement angolais." In addition, the author's sources indicate that when the notes were unfrozen, the IMF received word from the Paris Club that Russia had new claims on Angola, claims worth "about" $700 million— which suggests that the previous Sonangol payments totaling $774 million to Abalone had paid off the 16 notes in full, leaving $726 million outstanding, which Angola should still pay Abalone under the 1997 agreement.

23. See Global Witness, *Time for Transparency,* March 24, 2004, page 43. The memo was from 1999.

24. Under the escrow agreement with UBS, each Sonangol payment into Abalone should trigger UBS to pay Russia, which then authorized UBS to

send promissory notes back to Angola, progressively redeeming the old Soviet debt.

25. According to the text from the Swiss judicial authorities in December 2004, explaining the closure of the case. This text was reprinted in *Le nouvel observateur* on January 12, 2005, and can be found at http://www.afriquechos. ch/article.php3?id_article=578.

26. Angola sought to negotiate with Switzerland for this frozen money to return to Angola not as cash but as humanitarian aid. It is not immediately clear what happened to the funds.

27. Yossi Melman, "Me? An Oligarch?" *Haaretz*, March 18, 2005.

28. See "Un témoignage éclaire les dessous des ventes d'armes à l'Angola," *Le monde*, April 24, 2003.

29. "Alocucão pronunciada por Sua Excelência José Eduardo dos Santos Presidente da República de Angola, por ocasião da apresentacão de cartas credenciais do novo embaixador de França em Angola," *Angop*, February 23, 2001.

30. Michela Wrong, *In the Footsteps of Mr. Kurtz: Living on the Brink of Disaster in the Congo* (London: Fourth Estate, 2000), pages 294–297. Condensed from several sections of text.

31. Widely reported, from IMF, "Angola—Staff Report (Confidential) for the 2002 Article IV Consultation," March 18, 2002, page 17. See, for example, "Angola: Missing Oil Billions Unexplained," *IRIN news*, January 13, 2004.

32. Confidential document: Abalone Investments Ltd. at UBS, Geneva, May 23, 1997–Dec 31, 2000.

33. IMF, "Angola—Staff Report (Confidential) for the 2002 Article IV Consultation," March 18, 2002, Table 4, page 31.

34. Described in several places. See, for example, Anna Richardson, "Angola's oil boom fuels civil war," *The Independent on Sunday* (UK), 27 February 2000. The UN Human Development Report put Angola's under–five mortality rate in 2001 at 260 per 1,000 live births.

35. "País terá 200 mil casas em 3 anos," *Jornal de Angola*, June 8, 2006.

36. The late Christine Messiant.

37. See http://www.weforum.org/site/homepublic.nsf/Content/Global+Competitiveness+Program percent5CGlobal+Competitiveness+Report. Angola was not included in the 2005–2006 rankings.

38. Economist Intelligence Unit, Angola Country Report, first quarter 2005.

39. U.N. Human Development Report, Human Development Indicators, 2005.

40. IPEDEX, "Mapping Report—National Education Mapping and Training Needs in the Angolan Petroleum Industry 2001–2007," a confidential report for the Ministry of Petroleum/Norwegian Petroleum Directorate and Angola's Ministry of Education, Luanda, 2003.

41. "Estão a 'privatizar' o Presidente," *Semanário angolense*, June 6–13, 2006.

Chapter 10—Dokubo-Asari

1. See "Rebel Leader Threatens to Blow Up Gas Plant in Nigeria," VOA News, October 1, 2004.
2. See http://news.ft.com/cms/s/2f560d4a-f64b–11da-b09f–0000779e2340.html.
3. From BP, *Statistical Review of World Energy*, 2005. Government estimates have put gas reserves at 177 trillion cubic feet.
4. Widely reported. For example, http://www.cnn.com/SPECIALS/2000/democracy/bigger.picture/bush.gore/.
5. African Oil Policy Initiative Group, *African Oil: A Priority for U.S. National Security and African Development*, June 2002, available at http://www.iasps.org/strategic/africawhitepaper.pdf, pages 6 and 15.
6. Congressman William J. Jefferson, "Jefferson Participates in Congressional News Conference with African Oil Policy Initiative Group," June 12, 2002, available at http://www.house.gov/jefferson/press2002/pr_020612_oilpolicy.html.
7. http://www.alhajimujahiddokubo-asari.com/grandparents.htm.
8. Michael Ross, Department for International Development, *Drivers of Change.*
9. http://www.alhajimujahiddokubo-asari.com.
10. Osama Ibn Mujaheed Dokubo Asari, http://www.alhajimujahiddokubo-asari.com/children.htm.
11. Niger Delta News, undated, http://www.alhajimujahiddokubo-asari.com/niger_dinterv5.htm.
12. *Newswatch*, September 20, 2004.
13. Michael Peel, "Peace Accord Fails to Pour Oil on Troubled Waters of Niger Delta," *Financial Times*, December 22, 2004.
14. All Nigeria's oil production is in the Niger Delta, and oil made up $36.4 billion of $37.3 billion in exports in 2005; data from http://www.imf.org/external/pubs/ft/scr/2006/cr06180.pdf.
15. Information published by the Niger Delta Development Commission.
16. Interview with Asari, *Newswatch*, October 18, 2004.
17. See, for example, http://www.climatelaw.org/gas.flaring/report/gas.flaring.in.nigeria.html.
18. IMF, "IMF Concludes 2002 Article IV Consultation with Nigeria," January 2, 2003, available at http://www.imf.org/external/np/sec/pn/2003/pn0301.htm.
19. For example, http://www1.worldbank.org/prem/lessons1990s/chaps/Ctrynote7_AreNaturalResources.pdf; see also Paul Collier and Anke Hoeffler, *Greed and Grievance in Civil War* (Washington, D.C.: World Bank, 2000).
20. Eghosa Osaghae, *Crippled Giant: Nigeria since Independence* (London: Hurst, 1998), page 245.

21. http://www.dawodu.net/ogoni1.htm.

22. Karl Maier, *This House Has Fallen: Midnight in Nigeria* (New York: PublicAffairs, 2000), pages 88–89.

23. Ibid., page 90; Amnesty International, "Nigeria: Time to End Contempt for Human Rights," http://www.amnesty.org/ailib/intcam/nigeria/con4.htm; and Oronto Douglas and Ike Okonta, *Where Vultures Feast: Shell, Human Rights, and Oil in the Niger Delta* (San Francisco, Calif.: Sierra Club Books, 2001), pages 137–138.

24. One million naira.

25. Maier, *This House Has Fallen: Midnight in Nigeria,* pages 96–97.

26. Jedrzej Georg Frynas, *Oil in Nigeria: Conflict and Litigation between Oil Companies and Village Communities,* page 48.

27. Maier, *This House Has Fallen: Midnight in Nigeria,* page 103.

28. Georg Frynas, *Oil in Nigeria: Conflict and Litigation between Oil Companies and Village Communities,* page 54.

29. Maier, *This House Has Fallen: Midnight in Nigeria,* page 105, and http://www.hrw.org/reports/1999/nigeria/Nigew991–08.htm.

30. http://www.thisdayonline.com/archive/2001/01/24/20010124news04.html.

31. Douglas and Okonta, *Where Vultures Feast,* page 132, and Maier, *This House Has Fallen: Midnight in Nigeria,* page 107.

32. Ken Wiwa, *In the Shadow of a Saint* (Toronto: Knopf Canada, 2000), pages 139–140.

33. Widely reported; see, for example, Human Rights Watch, "Oil Companies Complicit in Nigerian Abuses," February 23, 1999, available at http://hrw.org/english/docs/1999/02/23/nigeri804.htm, and Human Rights Watch, "Nigeria: Crackdown in the Niger Delta," May 1999, available at http://www.hrw.org/reports/1999/nigeria2/Ngria993.htm#P36_586.

34. Also see a more detailed description of the attack in Christine Bustany and Daphne Wysham, Institute for Policy Studies, "Chevron's Alleged Human Rights Abuses in the Niger Delta and Involvement in Chad-Cameroon Pipeline Consortium Highlights Need for World Bank Human Rights Investment Screen," April 28, 2000, available at http://www.seen.org/PDFs/chevronfinal.doc, and a letter to Chevron Nigeria from Human Rights Watch at http://www.hrw.org/press/2003/04/nigeria040703chevron.htm.

35. Widely reported; see, for example, Human Rights Watch, "Letter to President Obasanjo," April 4, 2003, available at http://www.hrw.org/press/2003/04/nigeria040703obasanjo.htm.

36. Joseph A. Hurst-Croft, "The Prospects for Peace in the Niger Delta: Findings of a UK-Based Consultation," *Stakeholder Democracy Network* (London,) July, 2005.

37. Composed from telephone conversation with Stoddard and subsequent e-mail from him that described the events, July 2, 2006.

38. See Michael Peel, *Crisis in the Niger Delta: How Failures of Transparency and Accountability are Destroying the Region* (London: Chatham House), July 2005, page 3.

39. Newsom was referring to elections for around 300 political posts.

40. SPDC Working Paper, "Peace and Security in the Niger Delta: Conflict Expert Group Baseline Report," December 2003. Other estimates have put the figure closer to 200,000 barrels per day.

41. From the author's sources in the Delta.

42. http://www.irinnews.org/report.asp?ReportID=44956.

43. Interview with Australian Broadcasting Corporation, May 31, 2005, available at http://www.abc.net.au/foreign/content/2005/s1385109.htm.

44. http://www.onlinenigeria.com/links/Riversstateadv.asp?blurb=363.

45. The Nigeria Finance Ministry gives data on disbursements to states at http://www.fmf.gov.ng/detail.php?link=faac. The data used was the most recent at the time of writing, from April 2006.

46. See http://www.thisdayonline.com/nview.php?id=50453.

47. "Dokubo Rails at Obasanjo in Court," *Daily Sun* (Nigeria), April 13, 2006.

48. Niger Delta News, undated, http://www.alhajimujahiddokubo-asari.com/niger_dinterv5.htm.

49. See "Car Blast Near Nigerian Oil Port," BBC, April 30, 2006.

50. See "Nigeria Militants Tackle Ijaw Leaders Over Abuja Talks," *Vanguard*, April 11, 2006.

51. "Nigeria: Paradox of Life At Niger Delta Creeks," *ThisDay*, April 27, 2006.

52. See "As Oil Supplies Are Stretched, Rebels, Terrorists, Get New Clout," *Wall Street Journal*, April 10, 2006.

53. See "Asari Stops Attack Against Odili," *Elendu Reports*, July 18, 2006.

54. See Dino Mahtani, "Nigerian Militants Win Oil Drilling License," *Financial Times*, May 19, 2006.

55. SPDC Working Paper, "Peace and Security in the Niger Delta: Conflict Expert Group Baseline Report," December 2003.

56. See, for example, "Nigeria: Oil Giant Admits Aid Policies Helped Fuel Violence," *IRIN news*, May 4, 2005.

57. For a detailed analysis of this see Human Rights Watch, "Rivers and Blood: Guns, Oil, and Power in Nigeria's Rivers State," February 2005, available at http://hrw.org/backgrounder/africa/nigeria0205/.

58. http://hrw.org/reports/2006/nigeria0406/7.htm.

59. For a discussion of these issues, see "They Do Not Own This Place: Government Discrimination Against 'Non-Indigenes' in Nigeria," Human Rights Watch, April 2006.

60. See "Dream of Free Biafra Revives in Southeast Nigeria," *Reuters*, July 12, 2006.

61. www.africaresource.com/war/vol1.1/soyinka.html.
62. "We Have Crooks in Senate—Aliyu," *Online Nigeria*, April 21, 2005.
63. "24 Govs Face Trial Next Year, Says Ribadu," *ThisDay*, May 29, 2006.
64. See "Crisis in the Delta: How Failures of Transparency and Accountability Are Destroying the Region," by Michael Peel, Chatham House, July 2005.

Chapter 11—Global Witness

1. See *The Heart of the Matter: Sierra Leone, Diamonds & Human Security*, Partnership Africa Canada, (Ottawa), January 2000.
2. Matthew Hart, *Diamond: A Journey to the Heart of an Obsession* (Toronto: Viking, 2001). Quotation from diamonds.net and from Global Witness's 2004 annual report.
3. See *Angola Unravels*, Human Rights Watch (New York), September 1, 1999.
4. The effort was resuscitated only with the OECD antibribery convention of 1997, which Britain, France, and Germany did not even adopt until 1999 or 2000.
5. See World Bank, "Ten Things You Did Not Know about the World Bank and Anti-Corruption," http://web.worldbank.org/WBSITE/EXTERNAL/NEWS/0,contentMDK:20190202~menuPK:34457~pagePK:34370~piPK:34424~theSitePK:4607,00.html.
6. From a roundtable telephone conference that I participated in on July 16, 2004, at the launch of the addition of an anticorruption principle to the United Nations Global Compact.
7. See Nicholas Shaxson, "BP to Give Details of Angola Operations," *Financial Times*, February 2001.
8. See "All the Presidents' Men: the devastating story of oil and banking in angola's privatised war," Global Witness, March 2002.
9. From "A Dinosaur Still Hunting for Growth: Interview with Lee Raymond, ExxonMobil," *Financial Times*, March 12, 2002.
10. Other NGOs joined these discussions too.
11. Soros' Open Society Institute has, as already indicated, provided significant funding for this book.
12. Some payments, such as company tax payments on "profit oil," do flow from company to country.
13. IMF, "Angola—Staff Report (Confidential) for the 2002 Article IV Consultation," March 18, 2002, page 17.
14. Gabon, in its first stab at publishing data under EITI, omitted a category called "profit oil," which represents around half of government revenues. See http://www.finances.gouv.ga/eiti2.htm.
15. For example, the bits known as "cost oil" in "production sharing agreements," or PSAs. Under PSAs, oil revenues are divided between royalties (normally a

fixed per-barrel rate of payment to the government), and then the remainder is divided into "profit oil," which is shared between oil companies and the government according to a prenegotiated formula, and "cost oil," which constitutes the investment and running costs. The companies pay for the cost oil up front and then recoup their costs out of this portion of the oil revenue. Some oil contracts are not under production-sharing agreements but instead are under tax and royalty systems, which can produce similar end results but are structured differently.

16. The Catholic Church in Congo-Brazzaville has been more active than most on this issue, and the Nigerian EITI process has elicited some interest in the Nigerian press.

17. Global Witness has since revamped its methods to avoid repeating earlier errors. This has resulted in tighter standards of evidence, but probably at a cost of curbing how much data they feel confident about publishing.

Conclusion

1. Eva Joly, *Notre affaire à tous* (Paris: Arènes, 2000), page 159.

2. List of specified unlawful activities under U.S. anti-money-laundering laws, in Raymond W. Baker, *Capitalism's Achilles Heel: Dirty Money and How to Renew the Free-Market System* (Hoboken, N.J.: Wiley, 2005), page 187–188.

3. According to John Christensen of the Tax Justice Network, this term originates from someone once wrongly translating the word "heaven" instead of "haven."

4. http://www.taxjustice.net/cms/front_content.php?idcat=17.

5. See *Millennium Development Goals Report*, United Nations, 2006, page 22, which estimated global aid flows at $106 billion in 2005. http://unstats.un.org/unsd/mdg/Resources/Static/Products/Progress2006/MDG-Report2006.pdf.

6. http://www.electronic-economist.com/surveys/displaystory.cfm?story_id=E1_NSGJJV&CFID=83205164&CFTOKEN=1f9cb87-f852959e-c542–4621-a2d2–03cc9c7d99cd.

7. http://hsgac.senate.gov/110999_report.htm.

8. For example, http://www.guardian.co.uk/waronterror/story/0,566493,00.html.

9. Joly, *Notre affaire à tous*, page 164.

10. Raymond W. Baker, *Capitalism's Achilles Heel*, page 189.

11. See http://www.ft.com/cms/s/f61d9c12–1816–11db-b198–0000779e2340.html and http://www.ft.com/cms/s/cc4bcda2–20bb–11db-8b3e–0000779e2340.html.

12. "Tax Justice Network Sides with Europe's Tax Collectors," *Conservative Voice*, April 7, 2005, available at http://www.theconservativevoice.com/articles/article.html?storyid=4591.

13. William Snelgrave's book, *A New Account of the Slave Trade,* quoted in Baker, *Capitalism's Achilles Heel,* page 371.

14. Baker, *Capitalism's Achilles Heel,* page 48.

15. Senator Carl Levin, among others, suggested this one in the Omar Bongo Citibank Senate investigation, "U.S. Senate Permanent Subcommittee on Investigations Hearings on Private Banking and Money Laundering," November 9, 1999, page 104.

16. http://www.cato.org/pubs/pas/pa431.pdf.

17. I am paraphrasing Raymond Baker.

18. Baker, *Capitalism's Achilles Heel,* page 180.

19. See Martin Wolf, *Why Globalization Works* (New Haven, Conn.: Yale University Press, 2004), page 268; see his broad discussion of competitiveness on pages 249–277.

20. For example, News Corp (which owns Fox News) paid an effective 6 percent tax rate—including a zero tax rate in Britain—in some years, while corporate tax rates in its main markets have been above 30 percent. See "Rupert Laid Bare," *Economist,* March 18, 1999, available at http://www.economist.com/displaystory.cfm?story_id=319862. More recent estimates put News Corp's tax rates at 7 percent.

21. This is a quotation from ATTAC, http://www.attac.org/indexen/index.html.

22. IMF Working Paper by Xavier Sala-i-Martin and Arvind Subramanian, "Addressing the Natural Resource Curse: An Illustration from Nigeria," July 2003.

23. "Saving Iraq from its Oil," by Nancy Birdsall and Arvind Subramanian, *Foreign Affairs,* July/August 2004.

24. See, for example, http://www.economist.com/business/displaystory.cfm?story_id=5323394.

INDEX